Surveying the Courtroom

Surveying the Courtroom
A Land Expert's Guide to Evidence and Civil Procedure
Second Edition

John Briscoe

A Wiley-Interscience Publication
JOHN WILEY & SONS, INC.
New York - Chichester - Weinheim - Brisbane - Singapore - Toronto

Library of Congress Cataloging-in-Publication Data:

Briscoe, John.
   Surveying the courtroom: a primer for experts/John Briscoe. — 2nd ed.
     p. cm. — (Wiley series in surveying and boundary control)
   Includes index.
   ISBN 0-471-31840-X (alk. paper)
   1. Evidence. Expert—California. 2. Civil procedure—California. 3. Surveyors—Legal status, laws, etc.—California. I. Title. II. Series.
KFC1042.B75 1999
347.794′067—dc21

                                                 99-11597

10 9 8 7 6 5 4 3 2 1

To John and Katherine and Valerie

CONTENTS

CONTENTS

PREFACE

John Wiley & Sons and, in particular, Neil Levine of John Wiley & Sons are to blame for the exhumation of this book, first brought out by Roy Minnick's Landmark Enterprises in 1984. In short order the book's little run sold out. In recent years friends, calling from Brussels or Dublin or New York, have reported they had found a copy of the book in some second-hand bookstore. "What book?" I would ask.

So why a second edition? The law of evidence and procedure in land cases hasn't grown all that much different in 15 years. And I certainly haven't gotten wittier. As I wrote in the preceding paragraph, the blame has to be laid at the feet of John Wiley & Sons and Neil Levine: They thought too few people had read it.

Enough feigned beating of the breast. To the original Preface I have some emendations to make. I inexcusably omitted from the acknowledgments there Max Freeman of Stockton, California, one of the finest trial lawyers in the United States today and my earliest mentor, if you don't count my father, who died while I was in high school. I also omitted Robert Cathcart of the San Francisco Bar, a colleague of my father in Garret McEnerney's law offices in San Francisco before the Second World War. Like Max Freeman, Bob has been a great friend and mentor. And I also must acknowledge my partner and literary conscience David Ivester. It was he who came up with the title of this book 15 years ago and saved it from not a few embarrassments of error.

In the years since I wrote the first edition I have worked closely with Professors Steve Reisenfeld, Harry Scheiber, and David Caron at the Boalt Hall School of Law at the University of California at Berkeley. They have contributed immeasurably to my understanding of the law, and I want to thank them for that. I should also note that Tom Koester and Louis Claiborne, whom I acknowledged the first time around, are now members of my law firm, Koester in Alaska and Claiborne in, well, London, Nerja, San Francisco, and wherever else he happens to be.

Like the first edition, this one focuses on the rules of evidence and procedure found in the California and federal court systems. In the first edition I scarcely apologized for the seeming limitedness of this inclusionary policy, and so I thought I should venture forth a little here. Well, as it happens, I don't need to. While California continues to go its own way, hewing to its own rules of evidence, most other states have adopted the Federal Rules of Evidence. In fact, at last count 39 of the 50 have. They are Alabama, Alaska, Arizona, Arkansas, Colorado, Delaware, Florida, Hawaii, Idaho, Indiana, Iowa, Kentucky, Louisiana, Maine, Maryland, Michigan, Minnesota, Mississippi, Montana, Nebraska, Nevada, New Hampshire, New Mexico, North Carolina, North Dakota, Ohio, Oklahoma, Oregon, Rhode Island, South Carolina, South Dakota, Tennessee, Texas, Utah, Vermont, Washington, West Virginia, Wisconsin, and Wyoming.[1] Thus,

---

1. Charles E. Wagner, *Federal Rules of Evidence, Case Law Commentary* (Charlottesville: Lexis Law Publishing, 1997), pp. 881–900.

so far as the rules of evidence go, this little book thus covers at least 40 of the 50 states, plus the federal system.

In the Introduction to the first edition I wrote that I had two objectives. Since publication of that first edition 15 years ago, civil litigation has more and more come to be characterized by incivility, principally on the part of lawyers. It is a third objective of this book that it may contribute, however immeasurably, to the undoing of that deplorable state of affairs.

Finally, and most gratefully, I acknowledge the rigorous assistance of my indefatigable assistant, Theresa Fox. Had she not agreed to collaborate with me on this revision, it would not have happened.

San Francisco   John Briscoe
July 1998

PREFACE TO THE FIRST EDITION

This book was inspired, in a way of speaking, by a phalanx of professionals in a multitude of natural-resources fields, who have been my antagonists, my allies, my critics, colleagues and friends. The inspiration drew over a period of years, usually at late hours in stridently lit hotel rooms across the street from courthouses. On these occasions, some engineer or marine biologist whom, after sunup, I was to call to the witness stand would commence a cross-examination of me on the arcane and archaic legal process—the Byzantine ways of the hearsay rule, the mystical shifting of a burden of proof. Many of these questions I could not answer. And all of them kept me from the task at hand—preparing for the next day in court. It was on such occasions as these that I wished for a tidy book like this, that I could lend as I was ushering my expert to the door, with well-wishes for his sleep.

So in large measure, it is these experts in all their fields to whom this book is dedicated (some will say desiccated): To the land surveyors who have leveled me, the foresters who have treed me, the appraisers who have depreciated me, and the hydrologists who have deluged me—theirs are the questions I have tried to answer. As well, to lapse into gravity briefly, it is intended as the return of a favor. For much of my trade I have learned from them.

Much I have learned also from mentors such as N. Gregory Taylor, Judge Philip C. Jessup, and the late Jay L. Shavelson; from my law partner Edgar B. Washburn; from my colleague Alaska Assistant Attorney General G. Thomas Koester; and, in humbling ways, from adversaries such as Robert S. Daggett and Allan N. Littman of the San Francisco Bar, Deputy Solicitor General Louis F. Claiborne, and Michael W. Reed of the Department of Justice. For the indirect contributions of these men to the making of this book, I make this acknowledgment.

Acknowledgments are owing also to my persistent publisher Roy Minnick, and to my loyal and indefatigable assistants Fran Garland and Curt De Voe for their immense editorial efforts. To the extent this volume is devoid of violations of grammar, style and sense, it is their book.

Finally, it is *de rigeur* to lament the inability or neglect to mention the many others who are owed attribution. As for those, I take solace in the thought that the oversight likely offends them less than the acknowledgment might have.

San Francisco   John Briscoe
March 1984

Introduction

The expert in any of a number of technical fields is frequently, as a witness or as an adviser, drawn into the vortex of natural-resource litigation. And this expert, it seems, in short order comes to exhibit a markedly greater curiosity about the legal process than his counterparts in other kinds of law cases—physicians in personal-injury cases, for example. The expert in the land case may be a surveyor, title abstractor, property appraiser, engineer, geologist, hydrologist, oceanographer, biologist, or geographer. But there is one denominator that, in land cases, each of these experts bears in common with the other. It lies in the role each plays in a land or other natural-resource dispute, whether it concerns title, political or property boundaries, water rights, or environmental damage. In these cases the expert's role, for reasons this book may partly disclose, is much more like the lawyer's role than is the part of an expert in virtually any other kind of law case. It is perhaps on this account that the expert in a land case, more often than experts in other kinds of cases, asks to be initiated in the mysteries of the legal process. The general objective of this book is to do that.

The land expert's curiosity about the legal process is not comprehended with difficulty. He has been conscripted to play a high-stakes game whose rules, to the extent they are at all comprehensible, have as little apparent purpose as the elements of a secret-fraternity ritual. Furthermore, his questions are often rebuffed with the patronizing assurance of his lawyer that he needn't trouble himself with the incantations of hearsay and "shifting burdens." It is hoped that this book will help to dissipate much of the mystery that shrouds the rules of procedure and evidence in civil (i.e., non-criminal) cases. Two points of unpatronizing caution, however, are in order. First, these vast subjects, taught over a period of years in the law schools and mastered only with years more of practice, can be only superficially treated in a book of this kind. Second, it is one thing, say, to read Hoyle on bridge, and another to apprehend the subtleties of the game. This book's first objective, then, is to impart to the reader a nodding acquaintance with the principles of evidence and procedure.

A second objective is perhaps more ambitious. The law continually risks the loss of what public understanding and respect it enjoys; indeed, it risks the attainment of the public's contempt. It takes these risks perhaps not so much by unpopular decisions as by its use of involuted procedures, anachronisms, and in land cases even anachorisms.[1] To the nonlawyer the law may seem at times, particularly in the application of its rules of evidence and procedure, like a chess game, though having considerably more intricate rules and being played for dearer stakes.[2] (It does at times to the author.) Nevertheless, perhaps this book may also serve to show that the law, at least in the context of land and boundary cases, is a truly well-engineered instrument for the ascertainment of the facts and the just resolution of disputes. Churchill is reputed to have remarked that democracy is the worst form of government, save for all the others. With some understanding of it, perhaps our legal system can be similarly appreciated, and one can say with Melville, "[T]hus there seems a reason in all things, even in law."[3]

One observation may better than any other serve to justify this book as something more than a comparable volume for physicians, accountants, and other professionals who periodically encounter the judicial process. That observation, alluded to several paragraphs ago, concerns the unique role of the land expert (perhaps particularly the land surveyor), in relation to the law. It is perhaps best shown by contrast. A neurosurgeon does not excise a brain tumor by reference to a rule of law. Nor does an engineer design a bridge according to any laws (save those of physics). By contrast, the land expert, in making a survey or in forming an opinion as to the title, value, or location of property, applies law in the same way as a lawyer who is drafting deeds or subdivision restrictions. In turn, it might be added, the expert's professional ancestors often played a major role in shaping that law. President Thomas Jefferson, a land surveyor himself, persuaded Congress to establish the U.S. Coast Survey in 1807 and is reputed to have been the originator in 1785 of the rectangular system of surveys, from which has sprung many of the principles of boundary location.[4] (Presidents Washington and Lincoln were also land surveyors.) In the main, the law of boundaries and surveying has developed through government manuals and special instructions written by surveyors and through the customs and usages of surveyors, in time adopted by the courts as law. But again, the business of a physician or engineer is dictated not by law, but by a body of knowledge which has accumulated largely irrespective of the law. To be sure, it occasionally becomes law that, given certain symptoms in a patient, it is negligent of a physician not to order certain treatment. He will then be liable in law for any consequent injuries to the

---

1. One court has applied to *non-tidal* lakes and streams the ancient "tidelands trust" doctrine. *State of California v. Superior Court (Lyon)* 29 Cal.3d 210, 625 P.2d 239 (1981). Another court has ruled that the "ordinary high-water mark" boundary of tidal waters is to be determined by a mean of "neap high waters." *Teschemacher v. Thompson,* 18 Cal. 11, 79 Am. Dec. 151 (1861). According to the commentators, this latter ruling is a scientific and geographic impossibility. *See, e.g.,* Maloney and Ausness, The Use and Legal Significance of the Mean High-Water Line in Coastal Boundary Mapping, 53 N. Car. L. Rev. 185, 204 (1974).
2. "[T]o calculate is not in itself to analyze. A chess-player, for example, does the one, without effort at the other. It follows that the game of chess, in its effects upon mental character, is greatly misunderstood." Poe, "The Murders in the Rue Morgue," *The Complete Tales and Poems of Edgar Allan Poe* (New York: Modern Library, 1965), p. 141.
3. Herman, Melville, *Moby Dick* (New York: The Modern Library, 1950), p. 400.
4. 28 Journs. of Cong. 375–381 (1785). Jefferson was chairman of the committee appointed by the Continental Congress to prepare a plan for the survey and disposition of the public lands. Because of this, he is often mentioned as the inventor of the rectangular system. The prevailing opinion is that the system was not the result of any single individual's thinking. Patton, 1 Land Titles 289 (2d ed., 1957).

patient. That becomes the law because the careful physician would have ordered the treatment— that is his custom and practice. And when that medical practice is obsolete, the old rule falls.

A similar, continual modernization of rules cannot occur, however, in property law. The federal Constitution as well as many state constitutions provide that property may not be taken without due process of law, nor without the payment of just compensation. Any change in the law of title or boundaries will ineluctably enhance one estate (perhaps a neighbor's, perhaps the public's) and diminish another. But the owner of the second estate is constitutionally protected from this kind of diminution, or "taking." Therefore, it would seem that the law, cannot change the rules of property ownership, including the rules of boundaries, simply because the old rules are not seen to fit modern needs.

But it must have seemed that rules were changed when, for example, the United States Supreme Court held that the Illinois Legislature should not have conveyed certain lands within the bed of Lake Michigan to the Illinois Central Railroad and proceeded to invalidate the grant.[5] The Court's action in voiding the conveyance to the railroad may appear to have violated these constitutional provisions. The Court reasoned, however, that the Illinois legislature, having no authority to make the grant when it purported to do so, had in fact not conveyed the lands to the railroad at all. Thus, the Court was not effecting a change in the rules of property ownership, was it?

Similar cases in our history have invalidated patents to known mineral lands when they were found to have been issued in violation of the public land laws in force at the time of patenting.[6] It may be said in general that in such cases, had the grantee undertaken to learn the state of the applicable law when his conveyance was made, it should have come as no surprise that his patent was later held invalid. These cases may illustrate one facet of the concept that property may not be taken by the government without "due process of law." (Nor, of course, when taken for a public purpose, without just compensation.) The fact that the rules were ascertainable when the conveyance was made provided the person whose land was "taken" the due process of law the Constitution requires. These cases are better viewed, though, as ones in which no property was taken at all—with or without due process— because the conveyances were void when made.

Occasionally, however, and with seemingly increasing frequency, the rules of property ownership are changed, and invariably by court decisions and not by action of the legislature. These decisions either acknowledge straightforwardly that they are effecting a change of an old rule, or argue more or less persuasively that, to the perspicacious lawyer, the new rule was all along fully apparent in the older cases. These decisions have been made in response to public exigencies which, in the professed perception of the courts, cannot be met by other means. They are matters with which, affecting as they do the very property interests that are the subject of the land expert's profession, he ought to be somewhat acquainted. They may too serve to introduce one aspect of the judicial process—law-making.

Two not mutually exclusive kinds of lands in particular have been the subject of these recent decisions: (1) lands which underlie or abut bodies of water or are otherwise characterized as "wetlands" (such as marshes, swamps, etc.) and (2) accessways to those lands. These

---

5.   *Illinois v. Illinois Central Railroad,* 146 U.S. 387 (1892).
6.   *See, e.g., Morton v. Nebraska,* 88 U.S. (21 Wall.) 660 (1875), and *Saunders, Adm'x v. La Purisima etc. Co.,* 125 Cal. 159, 165 (1899).

lands are particularly suitable for wildlife preserves and public recreation, and they include submerged lands and lands on the shores of water bodies. Traditionally, government agencies acquired privately owned lands for wildlife sanctuaries and public parks by condemning them, an annoying process to the owner, but one which at least secured to him/her compensation for the taking of his property. Some recent cases, however, have held that such waterfront and submerged lands have in fact not been privately owned after all. (In these cases, there was "color" of private title, in the form of a government deed or patent.) Other decisions have taken a different approach—that the land, while nominally held in private ownership, is impressed with a "public trust" which delivers to the public such rights as recreation and fishing. What is left to the underlying fee owner is anyone's guess.

Several decisions of the California Supreme Court handed down in the 1980s are illustrative. In *City of Berkeley v. Superior Court*,[7] a 1980 decision, the court overruled a 1915 decision by the same court. The older decision had held that sales by the State in the nineteenth century of certain tidelands within San Francisco Bay passed full fee title to the purchaser, unencumbered by any retained state interest. In the *Berkeley* decision, the court affirmed that the present-day successor to the original purchaser of lands sold under this scheme—if the lands had been reclaimed by filling, the construction of levees, and so on—holds the full fee title. But, the court held that if the purchaser or his/her successors had not availed themselves of the opportunity to reclaim the lands by the date of the decision, the lands had become subject to a "public trust" easement in favor of the public for purposes traditionally described as "commerce, navigation, and fisheries."

Two decisions of the California Supreme Court the following year addressed the title to lands abutting non-tidal but navigable lakes and streams within California.[8] For approximately 100 years it had been the accepted rule in California that the owner of property bordering such a lake or stream owned to the low-water mark. This rule is found in opinions of California courts,[9] of the California Attorney General,[10] and in treatises on California real property law.[11] In the companion *Lyon* and *Fogerty* cases, the Supreme Court declined to accept the California Attorney General's suggestion that the title of such a riparian owner be limited to the high-water mark instead of the low-water mark. The Court did, however, accept the Attorney General's alternative submission that the lands lying between the low-water mark and the high-water mark be declared subject to a "public trust" akin to the tidelands public trust. No previous reported decision in California had intimated that these privately owned lands between high-and low-water marks were subject to such an interest in the State.

Still another decision of the California Supreme Court in this direction was the 1982 case of *City of Los Angeles v. Venice Peninsula Properties*.[12] There the Court imposed the

---

7. 26 Cal.3d 515 (1980).

8. *State of California v. Superior Court (Lyon)*, 29 Cal.3d 210, 625 P.2d 239 (1981); *State of California v. Superior Court (Fogerty)*, 29 Cal. 3d 240, 625 P.2d 256 (1981).

9. *See, e.g., City of Los Angeles v. Aitken*, 10 Cal.App.2d 460 (1935).

10. *See* 43 Ops.Cal.Atty.Gen. 291 (1964); 30 Ops. Cal. Atty. Gen. 262, 269 (1957); 23 Ops. Cal. Atty. Gen. 306, 307 (1954); 23 Ops. Cal. Atty. Gen. 97, 98 (1954).

11. *See, e.g.,* 1 A.G. Bowman, Ogden's Revised California Real Property Law § 15.3 1974).

12. 31 Cal.3d 288 (1982). The United States Supreme Court in March 1983 agreed to review this case, and revised it under the name *Summa Corporation v. State of California*, 466 U.S. 198, 80 L.Ed.2d 237 (1984).

public trust on asserted tidelands lying within the boundaries of lands granted by the Mexican government prior to California's annexation by the United States in 1848. The tideland-trust doctrine derives from the principle that, upon admission to the Union, each state received as an attribute of statehood title to the lands beneath navigable waters within its borders. This title, in early court decisions in California and elsewhere, had been said to be subject to an obligation—or public trust—on the part of the state to devote these lands for the public purposes of commerce, navigation, and fisheries. On the other hand, where such lands had been granted to private parties by prior sovereigns, it had been held that the state acquired no title to them. In California, property rights in lands granted to private parties by Spain and Mexico were protected both by customary principles of international law and by the Treaty of Guadalupe Hidalgo, which followed the Mexican War and effected the annexation of California by the United States. With respect to tidelands, this principle had been confirmed by the United States Supreme Court in a 1921 decision concerning property adjacent to North Island near San Diego; in that case, the Court held the State of California possessed no interest in such tidelands.[13] Thus, the *Venice Properties* case was squarely at odds with the position of the U.S. Supreme Court, which reversed the California Supreme Court in early 1984. (Interestingly, in a moment of unabashed candor, the California Supreme Court in its *Venice* decision noted that, without its imposing the public trust on the lands in controversy, the public would be unable to use them "without pursuing condemnation proceedings"—that is, without paying for them.[14]

In each of these three instances (*Berkeley, Lyon/Fogerty,* and *Venice Properties*) the private landowner was declared to have retained his or her fee simple title to the lands, and to be free to use them in any way that is consistent with the public-trust interest of the State.[15] Some discussion of the nature of this public-trust interest is warranted then, and not the least because other courts have employed approaches similar to that of California's.[16]

A first observation is that the public trust seems a desirable concept from a conservation standpoint; the California Supreme Court has ruled that it may be employed by the State to maintain lands in their natural condition for environmental studies.[17] Maintaining lands in their natural state, however, is but one of the many uses a state may make of lands subject to this interest. The cases from California hold that the easement may be used by members of the public to stroll, picnic, or even hunt on lands subject to the easement.[18] In addition, the State may dictate that these lands are to be used for harbor development or for hydrocarbon, geothermal, or mineral exploration and development, as it has with tidelands in Long Beach

---

13. *United States v. Coronado Beach Co.,* 255 U.S. 472 (1921).

14. 31 Cal.3d at 303.

15. *See, e.g., State of California v. Superior Court (Lyon), supra,* 29 Cal.3d at 232.

16. *See, e.g., O'Neill v. State Highway Department,* 235 A.2d 1 (N.J. 1967); *Seaway Co, v. Attorney General,* 375 S.W.2d 923, 929–930 (Tex. Civ. App. 1964) (public prescriptive rights in shoreline areas); *Palama v. Sheehan,* 440 P.2d 95, 97–98 (Hawaii 1968) (by virtue of ancient native use, royal patents granted to private landowners "ma ke kai" (along the sea) carried with them title only to the debris or vegetation lying on the shore), *Application of Ashford,* 440 P.2d 76 (Hawaii, 1968); and *State ex rel. Thornton v. Hay,* 462 P.2d 671, 679 (Ore. 1969) (declared public rights to all of the dry-sand areas of Oregon's beaches based on old English doctrine of custom.)

17. *State of California v. Superior Court (Lyon), supra,* 29 Cal.3d at 226, 234–235.

18. *Marks v. Whitney,* 6 Cal.3d 251, 259–260, 491 P.2d 374 (1971).

Harbor and along the southern California coast.[19] In fact, when all of the cases are gathered, it appears that there is little the State cannot do with its easement. A California Appeals Court wrote in 1974: "Proper 'trust purposes' have been held to include low level bridges of a freeway spanning a navigable waterway; a YMCA building for the welfare of merchant marine and servicemen of all branches of the armed forces; hunting water fowl from a boat; pleasure yacht harbors; keeping land in its natural state as a bird and wildlife sanctuary; and a civic center where the city spends other money for harbor improvements. The test appears to be more appropriately whether there is a general public purpose envisaged for use of the property which at least remotely benefits navigation, commerce or fisheries."[20] Thus it would seem that declaring property to be subject to such a "public trust" is tantamount to declaring that the landowner holds a mere "naked title to the soil." Indeed, that was the very expression used by the California Supreme Court in 1913 to characterize a private landowner's title to tidelands in San Pedro Bay.[21]

It bears emphasizing that court decisions that change the location of property boundaries, or that change the quality of property title, must be carefully constructed so as not to make that effect so plain. The rules must necessarily change over time, of course; it cannot be seriously suggested that there ought to be today absolute congruity with the rules of Blackstone's day. It is a matter of pragmatic judicial wisdom, however, that such changes must occur gradually, and with scarcely an appearance of change, for the sake of public confidence in the stability of the law. In addition, there are, as mentioned above, constitutional inhibitions to swift and unexpected changes in property law. A 1967 United States Supreme Court decision, *Hughes v. Washington*,[22] contains the following remarks by Justice Potter Stewart, concurring in the majority decision:

> As is so often the case when a State exercises its power to make law, or to regulate, or to pursue a public project, pre-existing property interests were impaired here without any calculated decision to deprive anyone of what he once owned. But the Constitution measures a taking of property not by what a State says, or by what it intends, but by what it does. Although the State in this case made no attempt to take the accreted lands by eminent domain, it achieved the same result by effecting a retroactive transformation of private into public property—without paying for the privilege of doing so. Because the Due Process Clause of the Fourteenth Amendment forbids such confiscation by a State, no less through its courts than through its legisla-

---

19. *Boone v. Kingsbury,* 206 Cal. 148, 273 P. 797 (1928); *Mallon v. City of Long Beach,* 44 Cal.2d 199, 282 P.2d 481 (1955); California Public Resources Code Section 6501-7062. (Since the *Boone* decision, the State has put hundreds of thousands of acres of tidelands along the California coast into oil production in the exercise of its tidelands interest.)

20. *San Diego Unified Port Dist. v. Coronado Towers, Inc.,* 40 A.C.A. 879. (Emphasis in original; the case was later "decertified" for publication, meaning it is not found among the officially reported California cases.)

21. *People v. California Fish Co.,* 166 Cal. 576, 598 (1913).

22. 389 U.S. 290, 298 (1967).

ture, and no less when a taking is unintended than when it is deliberate, I join in reversing the judgment.

For the court that would change the rules of property law, then, it is imperative to make it appear that, in the words of Shakespeare, "The law hath not been dead, though it hath slept."[23]

In the main, however, courts have been careful not to change the rules of real-property
law, particularly with respect to the boundaries of real property. The rule is familiar that corners once set, in the absence of gross error or fraud, control the boundaries of sections and
their subdivisions. With this principle the courts have not tampered. In addition, recent mani-
festation of this traditional attitude of courts toward stability in matters of boundaries and
title is found in the resolution of the dispute between the States of California and Nevada
over their common boundary. The case, being one between two states, was filed as an origi-
nal-jurisdiction matter in the United States Supreme Court, which then referred it to a special
master for the taking of evidence and testimony. The north–south boundary dividing the two
states is defined in the acts of Congress creating them as the 120th meridian of longitude
west of Greenwich, from the 42nd parallel of north latitude to the 39th parallel of north latitude (which intersection lies within Lake Tahoe). From there the boundary proceeds south-
easterly to a point on the Colorado River. Both the special master and the Supreme Court
rejected the suggestion that the line be resurveyed with the aid of modern, astronomical techniques, which testimony established as being more accurate than those employed by the original surveyors of the line.[24] Both the special master and the court were of the view that the
boundary should be held to lie where it had so long been presumed to be. And both noted
that if a new survey were ordered, it too might well in time be found erroneous with the development of even more precise surveying techniques.

Again, in a decision of the highest court of the State of New York, it was held that the
traditional method of locating the boundary of shorefront property should not be departed
from, notwithstanding arguments that new scientific discoveries made it possible to locate
the average high-water line more accurately than the methods traditionally employed by surveyors in the community. The Court's words reflect this more prevalent judicial inclination:

> Attaching real significance as we do to the importance of stability
> and predictability in matters involving title to real property, we hold
> that the location of the boundary to this shore-side property depends
> on a combination of the verbal formulation of the boundary line—i.e.,
> the high-water line—and the application of the traditional and custom-
> ary method by which that verbal formulation has been put in practice
> in the past to locate the boundary line along the shore. To accept the
> linguistic definition but then to employ an entirely new technique,
> however intellectually fascinating, for the application of that defini-
> tion, with the result that the on-the-site line would be significantly

---

23. Measure for Measure, Act 2, Scene 2.
24. Report of the Special Master, *California v. Nevada*, No. 73, Original, (1979), p. 38; *California v.
Nevada,* 447 U.S. 125 (1980).

differently located, would do violence to the expectations of the parties and introduce factors never reasonably within their contemplation. Thus, to recognize, as the town's argument must, that the type-of-grass test for location of the high-water mark may one day be replaced by an even more sophisticated and refined test for determining the high-water line, with a consequent shift again in the on-the-site location of a northern boundary line, is to introduce an element of uncertainty and unpredictability quite foreign to the law of conveyancing.

* * *

There was uncontroverted testimony here that it was the long-standing practice of surveyors in the Town of Southampton to locate shore-line boundaries by reference to the line of vegetation. To give effect to such uniform practice is not, as the town contends, to delegate arbitrary powers to surveyors to determine property lines; rather it is the obverse, namely, to recognize that property lines are fixed by reference to long-time surveying practice.[25]

The preceding discussion may give some sense of the enduring contribution of the land expert to the law—one which not even he can change in its fundamental respects when new technology emerges, much less when he has misgivings about the contribution of his professional forebears. This is not, however, to say that it is frivolous to ponder these fundamental respects of our system of property rights. It can be an enlightening and sometimes amusing endeavor. And occasionally a remark expresses such a tenet so plainly that, no matter how surely the reader's views had been in accord, he seems to find himself in a momentary lapse of faith. The Earl of Birkenhead made such a remark in relating the *Southern Rhodesia Land Case*, decided by the Judicial Committee of the Privy Council of Great Britain in 1918. The dispute concerned the ownership of the unoccupied lands in southern Rhodesia; the contestants were the British South Africa Company, the natives of southern Rhodesia, white settlers, and the Crown. As for the aborigines, Birkenhead wrote:

There was hardly a trace of land law. Probably it had never occurred to the native mind that land was capable of ownership. Whatever principle may be deduced from the decision of the King and his people and their subject tribes and the way in which they live, is after all an effort of the white man's brain.[26]

The decision in the case was for the Crown.

The reader who wishes to learn more of the theory and philosophy of law may wish to consult Holmes, *The Common Law* (1881), Plucknett, *A Concise History of the Common Law* (1929), and Wellman, *The Art of Cross-Examination* (4th ed., 1936). For a more thor-

---

25. *Dolphin Lane Assoc. v. Town of Southampton,* 372 N.Y.S.2d 52, 54 (1975).
26. Birkenhead, *Famous Trials of History* 213 (1926).

ough understanding of the points made in this book, he or she may wish to read some of the cases, statutes, and treatises cited in it.

* * *

I write occasionally of a rule of "procedural" as opposed to "substantive" law. Some examples should suffice to explain what I mean. Whether a document is filed on 13-inch-long paper or on 11-inch paper is clearly a matter of procedural law, as is the question of whether a case must be tried before six jurors or twelve. On the other hand, whether a trust can be devised so as to prevent forever the free conveying of the property of the trust is a matter of substantive law. This distinction, however, is not so clear as it may seem.[27] Is a statute of limitation, for example, which is often said to affect not the existence of a right but merely the ability to enforce it in court, a matter of procedural or substantive law? The answer to this question, which is beyond the scope of this book, is probably more problematic than ever as a result of remarks in a recent U.S. Supreme Court decision.[28]

The characterization of a rule as either procedural or substantive can assume enormous importance in a case; it arises principally in federal court cases in which the court's jurisdiction is predicated on the fact that the litigants are citizens of different states. This jurisdiction of a federal court is called "diversity of citizenship" or simply "diversity" jurisdiction. In such cases the federal court will employ its own rules of procedure, but will "borrow" the necessary rules of substantive law from the state in which the dispute arose. In land cases, this state is almost invariably the state in which the land is located. Is a statute of limitation, for example, which is often said not to affect the existence of a right but merely the ability to enforce it in court, a matter of procedural or substantive law? In California, for example, the period of limitations for bringing an action to recover the possession of real property is five years; in other states it is longer. And compounding this problem is the fact that there is no generally applicable statute of limitations for federal courts.

* * *

Part I of this book discusses those rules of evidence that seem to find frequent application in natural-resource cases. As much as possible, I have attempted to illustrate them with actual or hypothetical examples from such cases. Those rules of evidence that rarely find application in these kinds of cases (for example, the privilege against self-incrimination) are only mentioned in passing or wholly omitted for the sake of brevity. Part II describes the procedure followed in civil cases, from the filing of the lawsuit to its ultimate adjudication, perhaps by the

---

27. For cases dealing with whether a given rule is a procedural or substantive one, see *Wright, Miller and Cooper,* 19 Federal Practice and Procedure § 4504 *et seq.* (1982).

28. In *Block v. North Dakota,* 461 U.S. 273, 75 L.Ed.2d 840, 856 (1983), the Supreme Court discussed the effect of the twelve-year statute of limitations contained in the federal quiet-title statute, 28 U.S.C. § 2409a: "The statute limits the time in which a quiet title suit against the United States can be filed; but unlike an adverse possession provision, § 2409(f) does not purport to effectuate a transfer title. If a claimant has title to a disputed tract of land, he retains title even if his suit to quiet his title is deemed time-barred under § 2409a(f)."

United States Supreme Court. As to this seemingly inverted order of presentation—with the rules of evidence treated before the rules of procedure—I can only offer the tenuous explanation that the natural-resource expert may well be first interested in what evidence a court would permit him to introduce, and only second in the procedures of introducing that evidence at trial. Nevertheless, there is no overwhelming reason why this book must be read in the order presented.[29] The chapter on appeal, for example, describes the nature of our judicial system, for the purpose of illuminating differences between the trial courts and the appellate courts. Perhaps the reader will wish to read those pages first.

Finally, let me make some remarks about the material included in this book. There will be some subjects, perhaps the hearsay rule, that may seem to be treated in exasperatingly minute detail, while other subjects in which the reader is interested are altogether omitted. In this regard I can say only that I have been guided by what seemed to me to be the subjects most pertinent to natural-resource cases. There will be errors too, but in this case I don't suppose I can apologize for their existence by ascribing them, so to speak, to taste. And there will be other passages I will regret having written for yet another reason, when one is read with great glee by my opponent in court. For those occasions, I should like to reserve the right to assert that such passages fall in truth into the previous category.

---

29. "You can cut into Naked Lunch at any intersection . . . I have written many prefaces." William Burroughs, *Naked Lunch* 224 (1956).

Part I

The Rules of Evidence

Chapter 1

Relevance

"Some circumstantial evidence is very strong, as when you find a trout in the milk."

—Henry David Thoreau, Journal (November 11, 1854)

Relevance is the fundamental principle of the law of evidence. In a sense, all other rules of evidence in one manner or another address the question of relevance by focusing on the reliability or credibility of evidence, which are, on reflection, components of relevance. The less reliable or credible an item of evidence is, the less relevant it is. Professor McCormick wrote that relevance is logically the first rule of evidence,[1] and this principle has been codified in many jurisdictions. The federal rule provides:

> All relevant evidence is admissible, except as otherwise provided. . . . Evidence which is not relevant is not admissible.[2]

And many state rules are like California's:

> Except as otherwise provided by Statute, all relevant evidence is admissible.[3]

While the concept of relevance may seem self-evident, it is useful to examine the definitions that have been formulated. "Relevance," wrote Professor McCormick, "is probative worth."[4] The federal rule (for use in all federal and most California state courts) reads:

---

1. McCormick on Evidence § 151, at 314 (1954). (As with many texts, including perhaps this one, the author is more taken with the first edition than with subsequent ones. So there will be frequent citations to this first edition of Professor McCormick's classic work.)
2. Fed. R. Evid. 402.
3. Cal. Evid. Code § 351.
4. McCormick on Evidence § 151, at 314 (1954).

"Relevant evidence" means evidence having any tendency to make the existence of any fact that is of consequence to the determination of the action more probable or less prsbable than it would be without the evidence.[5]

The rule (used in all California courts) provides:

"Relevant evidence" means evidence, including evidence relevant to the credibility of a witness or hearsay declarant, having any tendency in reason to prove or disprove any disputed fact that is of consequence to the determination of the action."[6]

Both the California and the federal definitions of relevance, it should be noted, have two essential components: (1) The proferred evidence must tend to prove or disprove a proposition, and (2) that proposition must be of consequence to the action. Thus, if the location of a section corner, for example, is of no consequence to the action—that is, if its location would in no manner affect the outcome of the lawsuit—evidence of its location would be irrelevant. In addition, if the location of the corner were of consequence, but the offered evidence did not tend to establish its location, it too would be irrelevant.

These two components of the modern definition of relevance merge the older concepts of "material evidence" and "relevant evidence." Under the older terminology, evidence was "immaterial" if it was offered to prove a proposition that was of no consequence to the action; if the proposition were of consequence, but the proffered evidence tended neither to prove nor disprove it, the evidence was said to be "irrelevant."[7] Thus, an attorney's objection today that testimony is "irrelevant and immaterial" is redundant, because the modern definition of relevance comprises both relevance and materiality as those terms were formerly used.

The question of what matters are in issue in a case (and consequently material) is determined mainly by the pleadings and the rules of pleading, which are described in Chapter 10. It is also determined in large measure by the substantive law governing the dispute, which is explained briefly below, following the discussion of what might be called "exceptions" to the rule that relevant evidence is admissible.

While the general rule is that relevant evidence is admissible, the trial judge may in his discretion exclude evidence that is admittedly relevant (and not objectionable under the other rules of evidence) in certain well-established circumstances. These circumstances are where admitting the preferred evidence would cause (1) a danger of unduly prejudicing the jury, (2) a danger of confusing the jury, and (3) an undue consumption of the court's time.[8] Explicit photographs of a maiming victim are sometimes excluded, for example, even though relevant (to show the victim was in fact maimed), because the photographs may so inflame the jury that it loses sight of the more central issue—whether the defendant was in

---

5. Fed. R. Evid. 401. Nearly forty States have adopted the Federal Rules of Evidence. *See* Preface to the Second Edition.
6. Cal. Evid. Code § 210.
7. Model Evidence Code, Rule 1(8); 29 Cal. L. Rev. 190.
8. *See, e.g.,* Cal Evid. Code § 352; Fed. R. Evid. 403.

fact the attacker. For the same reason, a photograph showing the county surveyor arm-in-arm with a land developer who has been convicted of bribing public officials may be excluded. The photograph may arguably tend to show the dishonesty of the surveyor (the logical proposition is: An associate of a corrupt person is more likely to be corrupt than one who is not such an associate). But, its tenuous relevance may be outweighed by the danger of prejudice.

The rule that relevant evidence may be excluded if it might confuse the jury is infrequently invoked today. Modern trial judges are far more willing to allow complex scientific, economic, or other technical information to go to the jury than were their predecessors. A little more than half a century ago, judges were refusing to allow juries to study mortality tables in personal injury cases.[9] That is no longer the case.

Finally, relevant evidence may be excluded if it serves only to beat a dead horse. After three or four eyewitnesses to the stench of a sewage spill, the trial judge may refuse to allow more to take the stand. Furthermore, if an inordinate amount of time would be consumed establishing a minor or tangential point, one might expect the judge to draw the line. Justice Oliver Wendell Holmes, before he was seated on the U.S. Supreme Court, once pithily remarked that an objection that evidence would require too much court time "is a purely practical one—a concession to the shortness of life."[10]

It has occasionally been written that "unfair surprise" is an additional ground for excluding concededly relevant evidence. Rule 45(e) of the Uniform Rules of Evidence, for example, speaks of evidence whose admission will "unfairly and harmfully surprise a party who has not had reasonable opportunity to anticipate that such evidence would be offered." There is little case law on this subject, and there seems to be no great urgency for a rule excluding relevant evidence on this ground. Modern discovery procedures and the pretrial conference are largely designed to effect full disclosure of each party's evidence to the other. If the situation should arise, however, that notwithstanding diligent preparation one party is unfairly surprised at trial, the appropriate procedure would be for him to ask for a continuance to give him time to investigate the evidence and seek to counter it. For these considerations the drafters of California Evidence Code section 352 declined to include "unfair surprise" as a ground of exclusion.

In addition to the matters of prejudice, danger of confusion, and undue consumption of time, relevant evidence is typically excluded in other well-defined situations, for purposes of "public policy." The salutory effects of these rules ought to be obvious:

> 1. **Subsequent remedial conduct.** Most jurisdictions prohibit evidence that, following an accident for example, the defendant took steps to repair or cure the condition that assertedly caused the accident—fixing his automobile's brakes, mending a broken stairstep, seeing a psychiatrist (to cure a propensity to lash out upon insults to his ancestry). Were there not such a rule, remedial efforts after an injury would be discouraged.[11]

---

9.   *See McCaffrey v. Schwartz,* 285 Pa. 561, 576, 132 A. 810 (1926).
10.   *Reeve v. Dennett,* 145 Mass. 23, 11 N.E. 938, 943 (1887).
11.   *See* Fed. R. Evid. 407; Cal. Evid. Code § 1151.

2. **Compromises and offers to compromise.** Similarly, the law has an inveterate policy of encouraging litigants to pursue out-of-court settlements of their disputes. Consequently, no evidence of compromises made, or statements made in the course of negotiation, are admissible at trial. Should a party in a boundary dispute, for example, offer to compromise on a certain line mid-way between the two parties' assertions, and the offer is unaccepted, it is not admissible in a later trial of the dispute.[12]

3. **Existence of liability insurance.** For quite obvious reasons, the existence of liability insurance is inadmissible to prove the defendant's liability in the case—even assuming that the existence of the insurance is somehow an indication of his culpability.[13] Thought-provoking questions emerging from this rule include whether it covers the existence of title insurance in a case of disputed ownership or boundaries.

As mentioned above—and as ought to be apparent from this brief discussion of the concept of relevant evidence—what is relevant and, thus, admissible depends in large measure on the substantive rules governing the dispute. For example, in a lawsuit in which the sole issue is whether one party has acquired title to land by adverse possession, what evidence is relevant is a function of the elements needed to establish title by adverse possession in that state. If the payment of property taxes is not a necessary element (it is in a few states, such as California), then evidence of the payment of taxes is, strictly speaking, irrelevant. So also would be a "wild deed" purporting to convey the premises to the party asserting adverse possession, if color of title is not a necessary element.

It is beyond the scope of this book, however, to delve further into the substantive rules of land title and boundaries. The reader is referred to the latter part of Chapter 12, dealing with the trial of cases, where some of the obscure rules peculiar to the *trial* of land cases are briefly described. He will see how these "procedural" rules, too, can determine what is relevant evidence and what is irrelevant. For example, one such rule governing the trial of quiet-title actions (a rule which, strictly speaking, is not one of land title) provides that the plaintiff must prevail on the strength of his own title, and may not prevail merely by showing infirmities in defendant's title.[14] Thus, until the plaintiff in a title case has introduced sufficient evidence which, if unrebutted, would entitle him to a decree quieting his title, any evidence he may offer that is directed solely at attacking the defendant's claim is irrelevant. (As a matter of courtroom economy, the court may permit such evidence if the plaintiff's counsel promises forthwith to introduce evidence tending to establish plaintiff's title.)

A frequent source of relevance questions in land cases is the ambiguous document. The rules among the States vary greatly in the degree to which they permit "extrinsic" evidence—that is, evidence not found within the document itself—to be used to explain an unclear or ambiguous term or provision in a controlling document, such as a contract or deed.

---

12. *See, e.g.,* Fed. R. Evid. 408; Cal. Evid. Code § 1152, and § 1154.
13. *See, e.g.,* Fed. R. Evid. 411; Cal. Evid. Code § 1155.
14. *See, e.g., Ernie v. Trinity Lutheran Church,* 51 Cal.2d 702, 706 (1959).

Some courts, for example, hold that extrinsic evidence is admissible initially to establish an ambiguity in a deed. One leading California case has held the following:

> The test of admissiblity of extrinsic evidence to explain the meaning of a written instrument is not whether it appears to the court to be plain and not ambiguous on its face, but whether the offered evidence is relevant to prove a meaning to which the language of the instrument is reasonably susceptible.
>
> A rule that would limit the determination of the meaning of a written instrument to its four corners merely because it seems to the court to be clear and unambiguous would either deny the relevance of the intention of the parties or presuppose a degree of verbal precision and stability our language has not yet attained.

*   *   *

> Although extrinsic evidence is not admissible to add to, detract from, or vary the terms of a written contract, these terms must first be determined before it can be decided whether or not extrinsic evidence is being offered for a prohibited purpose. [Citations omitted.][15]

One final point will be appended to this first chapter on the principles of evidence. Should the judge wrongfully exclude, or wrongfully admit evidence, what is the effect of his error? Let it suffice for the moment to say that the error will not, on appeal, aid the party against whom the ruling went, unless he either objects or makes his "offer of proof." If inadmissible evidence is offered against his cause, it is incumbent upon a party to object to its introduction; if he neglects to object, he cannot on appeal predicate any reason for reversal on that ruling. Conversely, if properly admissible evidence is excluded by the trial judge, the party seeking to introduce it must make an "offer of proof." In the case of oral testimony, this is usually accomplished by having the witness, out of the jury's presence, tell what he would have testified to if allowed. Without this offer of proof, the reviewing court will have no way of knowing whether the excluded evidence was in fact properly admissible and, equally important, whether it would have made a difference in the outcome of the trial. The chapter dealing with the appeal of cases treats these points in more detail.

---

15. *P.G. & E. v. Thomas Drayage, Etc. Co.,* 69 Cal.2d 33, 37, 39 (1968). The *Thomas Drayage* case dealt specifically with a written contract, but the principle has been held applicable to deeds and other documents of conveyance as well. *Murphy Slough Assn. v. Avila,* 27 Cal.App.3d 649, 655 (1972); *Ferris v. Emmons,* 214 Cal. 501 (1931). It is often said that the "parol evidence" rule stated in the last-quoted paragraph is really not a rule of evidence, but one of contract law. *Estate of Gaines,* 15 Cal.2d 255, 264; 100 P.2d 1055 (1940).

Chapter 2

Documentary Evidence

"There is no such thing as a copy of a verbal [that is, oral] contract."

—*Martin v. Baines,* 217 Ala. 236, 116 So. 341 (1928)[1]

A.   AUTHENTICATION

Before a document may be received in evidence at trial, it must first be shown to be what the party offering it claims it to be. Authentication is the process of establishing this. If the field notes of a survey conducted by a witness are to be introduced, for example, the witness will be asked to identify them. This process is rather perfunctory (and accounts for much of the tedium at trial), but it is essential and cannot be ignored. The attorney will first take the field notes to the clerk and ask him to mark them with an exhibit designation, such as "Plaintiff's Exhibit 7" or "Defendant's Exhibit G." (Typically, the plaintiff's exhibits are assigned numbers, and defendant's exhibits are given letters of the alphabet.) This procedure is called "marking for identification," and it merely provides a shorthand way of referring to the document so that it will be clear from the reporter's record of the trial what document is being discussed. Marking the document for identification, however, is not the same as offering it into evidence, which occurs later.

Next the attorney will show the document to the witness, and examine him or her along these lines:

> Q.   Mr. Bailey, I show you Plaintiff's Exhibit 7 for identification and ask if you can identify this document for the Court?
>
> A.   Yes I can. These are the original field notes of a survey I made of the subject property on May 7th and 8th of last year.

---

1.   *Martin v. Baines* was decided, obviously, before the proliferation of recording devices. Today, I suppose, there could be audio and video "copies" of oral contracts.

Q.   When were the entries in these notes made?

A.   They were made as I was making the survey in the field, sir. They are not transcriptions made later in my office.

In this simple instance the document could probably be received in evidence at this point. It has been authenticated, it is no doubt conceded to be relevant, and the hearsay problem (discussed below) has been overcome. At this point the attorney might say:

Counsel: Your Honor, I offer Plaintiff's Exhibit 7 into evidence.

The Court: Are there any objections?

Opposing Counsel: No, your Honor.

The Court: Then it will be received in evidence.

From this point forward, the document will be referred to as "Plaintiff's Exhibit 7 in evidence," rather than "Plaintiff's Exhibit 7 for identification." It is good practice for someone on the litigation team (a junior lawyer, a paralegal, or perhaps the surveyor or title abstractor, while he is not on the witness stand) to keep a record of which documents have been marked for identification, and which have been received in evidence.

The process of authentication does not always proceed so smoothly. On cross-examination, or on "voir dire" examination,[2] the lack of personal knowledge of the witness might be shown through the questioning of the opposing attorney:

Q.   You have identified Defendant's Exhibit M as being the field notes of a survey conducted in 1963 by a Mr. Newley. Did you know Mr. Newley personally?

A.   No.

Q.   Did you ever do business with Mr. Newley's firm?

A.   No.

Q.   Aside from what you may think you have learned from these purported field notes, do you have any personal knowledge that Mr. Newley in fact ever conducted such a survey?

A.   No.

Q.   Are you aware that Mr. Newley was stripped of his land surveyor's license for charging for surveys that were never performed?

A.   No.

At this point, additional evidence that Defendant's Exhibit M is in fact the field notes of the survey in question would no doubt need to be introduced before the exhibit could be received into evidence.

---

2.   "Voir dire" examination, in this context, is the questioning of a witness by an attorney other than the one who called the witness, prior to the witness's completing his direct testimony. It is done before the witness testifies to the critical point of his testimony—e.g., his observation of an assault—to determine, in this example, whether he actually observed what he is about to describe.

The California definition of authentication is this:

> Authentication of a writing means (a) the introduction of evidence sufficient to sustain a finding that *it is the writing that the proponent of the evidence claims it is* or (b) the establishment of such facts by any other means provided by law.[3]

Thus, even though a writing is relevant and not subject to any exclusionary rule, a foundation must be laid by authentication before it can be introduced into evidence.[4]

The federal rule respecting authentication, Rule 901 of the Federal Rules of Evidence, is interesting for its use of examples of how matter may be authenticated:

> (a) *General provision.* The requirement of authentication or identification as a condition precedent to admissibility is satisfied by evidence sufficient to support a finding that the matter in question is what its proponent claims.
>
> (b) *Illustrations.* By way of illustration only, and not by way of limitation, the following are examples of authentication or identification conforming with the requirements of this rule:
>
> (1) *Testimony of witness with knowledge.* Testimony that a matter is what it is claimed to be.
>
> (2) *Nonexpert opinion on handwriting.* Nonexpert opinion as to the genuineness of handwriting, based upon familiarity not acquired for purposes of the litigation.
>
> (3) *Comparison by trier or expert witness.* Comparison by the trier of fact or by expert witnesses with specimens which have been authenticated.
>
> (4) *Distinctive characteristics and the like.* Appearance, contents, substance, internal patterns, or other distinctive characteristics, taken in conjunction with circumstances.
>
> * * *
>
> (6) *Telephone conversations.* Telephone conversations, by evidence that a call was made to the number assigned at the time by the telephone company to a particular person or business, if (A) in the case of a person, circumstances, including self-identification, show the person answering to be the one called, or (B) in the case of a business, the call was made to a place of business and the conversation related to business reasonably transacted over the telephone.

---

3. Cal. Evid. Code §1400 (emphasis added).
4. Cal. Evid. Code § 1401 (a).

(7) *Public records or reports.* Evidence that a writing authorized by law to be recorded or filed and in fact recorded or filed in a public office, or a purported public record, report, statement, or data compilation, in any form, is from the public office where items of this nature are kept.

(8) *Ancient documents or data compilation.* Evidence that a document or data compilation, in any form, (A) is in such condition as to create no suspicion concerning its authenticity, (B) was in a place where it, if authentic, would likely be, and (C) has been in existence 20 years or more at the time it is offered.

* * *

The Federal Rules of Evidence, it should be remembered, are also those of nearly 40 states. Similarly permissible methods of authentication are found in the laws of other jurisdictions, such as California.[5]

Presumptions, which are considered in a later chapter, frequently help in authenticating documents. In California, for example, a "book, purporting to be printed or published by public authority, is presumed to have been so printed or published."[6] Similarly, in common with many other states, California law presumes the authenticity of "ancient documents," providing four conditions are met. The document must be at least 30 years old; it must be in such a condition that raises no suspicion concerning its authenticity; it must have been kept or found where one would expect it to be; and lastly, it must have been generally treated as authentic by persons having an interest in the matter.[7] Another presumption that can aid in authenticating a document holds that a writing is presumed to have been truly dated.[8]

All such presumptions respecting the authenticity of documents, however, may be rebutted by showing the document is fake or forged. These presumptions are not conclusive, but their effect is that unless contrary evidence is shown, the document will be presumed authentic.

In the case of government documents, the process of authentication is greatly hastened by the use of certified copies. (Two other evidentiary hurdles to the introduction of documents—the best evidence rule and part of the hearsay rule—are also overcome by the use of certified copies.) Virtually all states, and federal law, provide that properly certified copies are "prima facie" evidence of (a) the fact that the original is on file with the public entity, and (b) what the original contains. By "prima facie" it is meant that the opposing party may seek to show that there is no original, or that the original is not genuine, or that the purported copy is not a true copy, but if he or she cannot, the certification establishes the authenticity of the document. Thus a certified copy of the field notes of a government survey will be treated as authentic as if one produced the original and authenticated it. It should be apparent then that using a certified copy is preferable even to using the original. Assuming one could obtain an original government document, it would still be necessary to establish its authenticity, which

---

5. *See, e.g.,* Cal. Evid. Code §§ 1410, 1413, 1415, 1416.
6. Cal. Evid. Code § 644.
7. Cal. Evid. Code § 643.
8. Cal. Evid. Code § 640.

usually is done by calling as a witness the custodian of the document. Using a certified copy dispenses with the need to call a "live" witness for authentication. (Certification of documents is treated in greater detail in Section B of this chapter.)

A few comments about the authentication of photographs should be made. The principle that applies to documents also applies to photographs; the photograph must be shown to depict what its proponent claims it depicts. An aerial photograph of the property whose boundaries are disputed can be authenticated by the testimony of a witness who is sufficiently familiar with the property that he can recognize it from the photograph. It is not necessary to call the photographer to the stand, although that is one very common manner of authenticating photographs.

Many government entities today take aerial photographs and make the negative and prints available to the public. Since, as with other government documents, these photographs can be certified by the agencies, for purposes of authentication they may be preferable to photography taken by private firms. When obtaining certified copies of government photographs, it is advisable to ask that the certificate specify as much information about the photograph as possible. Detailed information is less important when the photograph is to be used simply as an illustration, to orient the judge or jury. But frequently such photographs are used as a basis for drawing inferences. In a river boundary case, a series of photographs showing the river's location at various times over a period of years may help to establish whether movements of the river were gradual and imperceptible (accretion and erosion) or sudden and perceptible (avulsion). In such a case, one would want to know as much information about the photographs as possible, including the date and time the picture was taken, possibly the kind of camera and size of lens, and so on. If this information can be included on the certificate of genuineness, so much the better.

In addition to the presumptions discussed, there are other commonplace methods of establishing the authenticity or genuineness of a document. A common method of proof is a comparison of the offered writing with some other admittedly genuine writing, called an "exemplar." The comparison may, of course, be made by an expert witness in order to form his opinion.[9] But the two writings—the disputed and the exemplar—may be offered in evidence and considered by the jury or other trier of fact as circumstantial evidence to prove genuineness or lack of genuineness.[10] The foundation for such evidence is proof of the preliminary fact of genuineness of the exemplar.

The court must find that it was "admitted or treated as genuine" by the adverse party, or it must be "otherwise proved to be genuine to the satisfaction of the court."[11] However, California Evidence Code section 1419 relaxes the requirement for proof of ancient writings:

> Where a writing whose genuineness is sought to be proved is
> *more than 30 years old,* the comparison under Section 1417 or 1418
> may be made with writing purporting to be genuine, and *generally*

---

9.   Cal. Evid. Code § 1418.
10.   Cal. Evid. Code § 1417; *Castor v. Bernstein,* 2 Cal.App. 703, 705, 84 P. 244 (1906); *People v. Storke,* 128 CT1. 486, 488, 60 P. 1090 (1900); *People v. Gaines,* 1 Cal.2d 110, 115, 34 P.2d 146 (1934); *People v. Davis,* 65 Cal.App.2d 255, 257, 150 P.2d 474 (1944) [prosecution for maintaining illegal betting establishment; comparison of questioned betting marker with admittedly genuine loan application].
11.   Cal. Evid. Code § 1417.

> *respected and acted upon as such,* by persons having an interest in
> knowing whether it is genuine.

The exemplar must be one which was written naturally and independently of the purposes of the litigation. A specimen writing or signature prepared specially for the use of an expert (*"Post litem motam"*) is not admissible. And, a defendant cannot offer exemplars of his handwriting, made at the trial or after his arrest, because of the possibility of fraud in allowing him to corroborate his own testimony by a prepared specimen.[12]

If a letter or telegram is sent to a person and a reply is received in due course purporting to come from that person, this is sufficient evidence of genuineness.[13] California Evidence Code section 1420 codifies this rule as follows: "A writing may be authenticated by evidence that the writing was received in response to a communication sent to the person who is claimed by the proponent of the evidence to be the author of the writing."

The authenticity of a letter may appear from its contents, as where it states facts which would be known to the person whose letter it purports to be, or appears to be part of a series of letters between two parties. "If its tenor, subject-matter, and the parties between whom it purports to have passed make it fairly fit into an admitted or proved course of correspondence and constitute an evident connecting link or part thereof, these circumstances justify its admission."[14] In California, the rule is as follows:

> A writing may be authenticated by evidence that the writing
> refers to or states matters that are unlikely to be known to anyone
> other than the person who is claimed by the proponent of the evidence to be the author of the writing."[15]

The authenticity of an instrument relied on by the plaintiff or defendant also may be established prior to the trial, in discovery proceedings, or at a pretrial conference. These matters are considered in Chapters 11 and 12, below.

Writings that have been plainly altered present a special situation. The California statute provides the following:

> The party producing a writing as genuine which has been altered, or appears to have been altered, after its execution, in a part *material to the* question in dispute, *must account for the alteration or appearance* thereof. . . . If he does that, he may give the writing in evidence, but not otherwise.[16]

California's statute goes on to specify the kinds of showing that will adequately account for the alteration:

---

12. *People v. Briggs,* 117 Cal.App. 708, 711, 4 P.2d 593 (1931).
13. 9 A.L.R. 989 (1920); 52 A.L.R. 583 (1928).
14. *Chaplin v. Sullivan,* 67 Cal.App.2d 728, 734, 155 P.2d 368 (1945).
15. Cal. Evid. Code § 1421.
16. Cal. Evid. Code § 14021 *see King v. Tarabino,* 53 Cal.App. 157, 165, 199 P. 890 (1921).

1. That the alteration "was made by another, without his concurrence."
2. That it "was made with the consent of the parties affected by it."
3. That it "was otherwise properly or innocently made."
4. That it "did not change the meaning or language of the instrument."[17]

B.   THE "BEST-EVIDENCE" OR "ORIGINAL-DOCUMENT" RULE

The best-evidence rule may be the most egregiously misnamed rule in the law. Prosecutors may speak of the smoking gun as the "best evidence" of guilt, and surveyors may debate what constitutes the best evidence of an obliterated corner. But the best-evidence rule is not a rule of general application that admits only the most probative or persuasive evidence. It applies only to documents, and it would be more aptly called the "original-document" rule, which is the name that will be used here henceforth and hereafter.

Succinctly stated, the rule is this: In proving the terms of a writing, when the terms are material to the case, the original writing must be produced unless it is shown to be unavailable for some reason other than the serious fault of the one who wishes to prove the writing's terms.[18]

The purpose of the rule is plain. If a question arose at trial concerning an entry in the field notes of an original government survey, the most reliable evidence of what was entered would be the original field notes themselves. A handwritten or typewritten transcription of the original notes—not to mention oral testimony of their contents—would be too susceptible to errors, whether inadvertent or purposeful. This problem is presented less by modern photocopying methods, but nonetheless persists.

> The reasons for the rule are thus stated in [section 602 of] the Model [Evidence] Code: "Slight differences in written words or other symbols may make vast differences in meaning; there is great danger of inaccurate observation of such symbols, especially if they are substantially similar to the eye. Consequently there is opportunity for fraud and likelihood of mistake in proof of the content of a writing unless the writing itself is produced. Hence it should be produced if available.[19]

The statement of the rule given above, it should be noticed, specifies that the original is required only when the terms of the writing are material to the case. Stated in reverse fashion, if the writing is "collateral" to the principal issues of the case, the original need not be produced. Suppose a witness testifies that he first observed the witness tree on December 4. Asked how he knows it was December 4, he replies that he was carrying that day's newspaper at the time and made a mental note of the date given below the paper's banner. In such an instance, it would hardly be necessary to produce the original of the newspaper to prove the date.

---

17. Cal. Evid. Code § 1402.
18. McCormick on Evidence § 196, at 409 (1954).
19. Model Code of Evcidence rule 602, comment at 300 (1942); *see also* McCormick on Evcidence § 197, at 410 (1954); 4 Wigmore on Evidence § 1179 (rev. ed. 1974); Cal. Evid. Code § 1500, comment of California Law Revision Commission.

The California rule reads:

> Except as otherwise provided by statute, no evidence other than the original of a writing is admissible to prove the content of a writing.[20]

A videotape is a writing under Section 250.[21] So is a tape recording,[22] as well as a magnetic tape containing computer data entries.[23]

The federal rule is distinctive in one significant respect. It is not called the "best-evidence rule" but rather "Requirement of Original." Like the California rule, the federal rule has been expanded to cover not only writings, but also recordings and photographs.[24]

The original-document rule comes into operation when an attempt is made to offer *secondary* evidence (e.g., a copy or oral testimony) to prove the contents of an original writing. Sometimes, however, it may be difficult to determine whether a writing is an original document or secondary evidence for purposes of the rule. If there is in a practical sense no single original but several duplicates of equal reliability, any one could be considered primary evidence, and the policy of the rule would be satisfied. This theory is approved by modern authorities, but the cases still make conflicting and confusing distinctions between different kinds of duplicates.[25] Carbon copies—rarely generated in the latter-day commercial world—were generally regarded as duplicate originals and may be introduced without showing the unavailability of the original.[26] Printed and other mechanically reproduced copies should, like carbons, be regarded as originals for purposes of the rule. However, except as to business records under the California statute, photographs and photocopies are still generally considered secondary evidence, unless the copies were made in the regular course of business.[27]

Generally, when an original writing is shown to be unavailable for some reason other than the serious fault of the proponent—that is, the party offering the evidence—then such

---

20. Cal. Evid. Code § 1500.
California Evidence Code section 250 in turn defines "writing" to include "handwriting, typewriting, printing, photostating, photographing, and every other means of recording upon any tangible thing any form of communication or representation including letters, words, pictures, sounds, or symbols or combinations thereof."
The exception for "collateral writings" is found in Evidence Code section 1504, which makes a copy admissible "if the writing is not closely related to the controlling issues and it would be inexpedient to require its original."

21. *Rubio v. Superior Court,* 202 Cal.App.3d 1343, 249 Cal.Rptr. 419 (1988).

22. *O'Laskey v. Sortino,* 224 Cal.App.3d 241, 273 Cal.Rptr. 674 (1990).

23. *Aguimatang v. California State Lottery* 234 Cal.App.3d 769, 286 Cal.Rptr. 57 (1991).

24. The general rule is Federal Rule of Evidence 1002; the exception for collateral writings, recordings or photographs is contained in Rule 1004(4).

25. *See* 64 Harv.L. Rev. 1369 (1951); McCormick on Evidence § 206, at 419 (1954); 4 Wigmore on Evidence § 1232 (rev. ed. 1974).

26. *Pratt v. Phelps,* 23 Cal.App. 755, 757, 139 P. 906 (1914); *Edmunds v. Atchison etc. Ry. Co.,* 174 Cal. 246, 247, 162 P. 1038 (1917); *Hughes v. Pac. Wharf Co.,* 188 Cal. 210, 219, 205 P. 105 (1922); *People v. Lockhart,* 200 Cal.App.2d 862, 871, 19 C.R. 719 (1962).

27. Cal. Evid. Code §§ 1500, 1550; *See Hopkins v. Hopkins,* 157 Cal.App.2d 313, 321, 320 P.2d. 918 (1958); Annot. 142 A.L.R. 1270 (1942); Annot. 76 A.L.R.2d 1356 (1958).

evidence is admissible. If the original is shown to have been lost or destroyed, for example, secondary evidence of its contents may be introduced.[28] The same is true if the original cannot be procured through the court's subpoena power or other available means, or if it is in the hands of the opponent.[29] Thus, in preparing for trial, careful record should be kept of the location and custodian of the originals of all pertinent documents. If an original cannot be located, the efforts made to find it should be meticulously recorded.

It is generally enough to show the exhaustion of ordinary sources of information and means of discovery; the matter, however, is largely committed to the discretion of the trial judge. "The general rule concerning proof of a lost instrument is, that reasonable search shall be made for it in the place where it was last known to have been, and, if such search does not discover it, then inquiry should be made of persons most likely to have its custody, or who have some reason to know of its whereabouts."[30]

In *People v. Jackson*,[31] a trial for conspiracy to violate the Corporate Securities Act, prosecution witnesses testified to the contents of the books of two corporations after a showing that the books had last been seen in the possession of defendants, and counsel for defendants in open court had denied such possession. The court said that for all practical purposes the books "were as effectively lost as though their actual loss or destruction had been conclusively established."

California Evidence Code section 1501, based on Uniform Rule 70(1)(a), makes explicit what was only implied in the former California statute and recognized by case law: The loss or destruction must have been "without fraudulent intent on the part of the proponent of the evidence."[32]

Writings are often intentionally destroyed for wholly innocent reasons unconnected with any contemplation of litigation. Proof of such circumstances is a sufficient foundation for offer of secondary evidence. Thus, in *Guardianship of Levy*[33] the respondent sought to prove that a child's deceased mother had expressed a preference for respondent as guardian of the child in letters written to her. She testified that the letters were received in the Philippines where her husband was on military duty but were destroyed when she prepared to return to the United States, because of baggage limitations. The trial judge held that this explanation was a sufficient foundation for her oral testimony of the contents of the letters.

Proponents of secondary evidence cannot always count on a court to accept this theory, however. In *People v. King*,[34] police officers took tape recordings of conversations with defendants, transcribed them on discs, and then, pursuant to their usual routine, erased the tapes

---

28.   Cal. Evid. Code § 1501; Fed. R. Evid. 1004(1).

29.   Fed. R. Evid. 1004(2) and (3); Cal. Evid. Code §§ 1502, 1503.

30.   *Kenniff v. Caulfield*, 140 C. 34, 40, 73 P. 803 (1903); *see also Ulm v. Prather*, 49 Cal.App. 141, 144, 192 P. 878 (1920); *Cotton v. Hudson*, 42 Cal.App.2d 812, 814, 110 P.2d 70 (1941) ["The exactness of proof may be relaxed in proportion to the evidentiary weight or value of the instrument, with due consideration given to the interest in its production of the party offering to prove loss"]; *Cheek v. Whiston*, 159 Cal.App.2d 472, 477, 323 P.2d 1028 (1958) [sufficiency of foundation is within discretion of trial judge].

31.   24 Cal.App.2d 182, 198, 74 P.2d 1085 (1937).

32.   *See Hiberna etc. Soc. V. Boyd*, 155 Cal. 193, 199, 100 P. 239 (1909) [judgment in foreclosure suit, destroyed in San Francisco fire of 1906].

33.   137 Cal.App.2d 237, 249 (1955).

34.   101 Cal.App.2d 500, 507, 225 P.2d 950 (1950).

for further use. Thus, the original "writing" was intentionally destroyed, but in good faith after obtaining a faithful reproduction. Nevertheless, the court declared that the discs would not be received.

Miscellaneous statutes are frequently found relating specifically to property records and providing relief from the original-document rule. As examples:

a. *Official Record of Document Affecting Property Interest.*   The official authorized record of "a document purporting to establish or affect an interest in property" is prima facie evidence of the existence and content of the original recorded document "and its execution and delivery by each person by whom it purports to have been executed."[35]

b. *Recital in Mineral Patent.*   The recital in a federal patent for mineral lands within California of the date of location of the claim is prima facie evidence of that date.[36]

c. *Deed by Officer in Pursuance of Court Process.*   (1) A deed purporting to have been executed by a proper officer under legal process of a court of record of this state, acknowledged and recorded in the county where the real property is situated, or (2) the record of such deed, or a certified copy of such record, is prima facie evidence that the property or interest described "was thereby conveyed to the grantee named in such deed."[37]

d. *Certificate of Purchase or of Location of Lands.*   A certificate of purchase or location of lands in this state, issued pursuant to federal or California law, is prima facie evidence that the holder or assignee is the owner of the described land. However, this presumption may be overcome by proof that, at the time of the location or of filing a preemption claim, (1) the land was in the adverse possession of the opponent of those under whom he claims or (2) the opponent is holding the land for mining purposes.[38]

e. *Authenticated Spanish and Mexican Title Records.*   Duplicate copies and authenticated translations of original Spanish title papers, prepared and authenticated pursuant to an old statute, are admissible as prima facie evidence with the effect of originals and without proving execution of the originals.[39]

The original document rule is inapplicable if the writing "was not reasonably procurable by the proponent by use of the court's process or by other available means."[40] Prior California cases reached the same result by treating the writing as "lost."[41]

Even though a writing is outside the reach of process, the original-document rule applies if other means of procurement are available. But where the writing is in the hands of a third person whose attitude is hostile and who refuses to give it up, some authorities hold that compliance is excused. [42]

---

35.   Cal. Evid. Code § 1600.
36.   *See* Calif. Pub. Res. Code § 2325.
37.   Cal. Evid. Code § 1603.
38.   Cal. Evid. Code § 1604.
39.   Cal. Evid. Code § 1605.
40.   Cal. Evid. Code § 1502.
41.   *Zellerbach v. Allenberg,* 99 Cal. 57, 73, 33 P. 786 (1893); cf. *Heinz v. Heinz,* 73 Cal.App.2d 61, 66, 165 P.2d 967 (1946) [original negatives and prints in photographer's studio in New York].
42.   McCormick on Evidence § 202, at 414 (1954); 4 Wigmore on Evidence § 1211 (rev. ed. 1974).

A special type of impracticability occurs where the writing is in the control of the adverse party. If the proponent wishes to inspect or introduce the original in evidence, he may employ the discovery procedure of inspection or serve a subpena duces tecum. But if he merely wishes to lay a foundation for the introduction of secondary evidence of its contents, he may give the adverse party a simple notice to produce it, which need not be in any particular form and may be oral or written.

Thus, California Evidence Code section 1503 makes a copy admissible "if, at a time when the writing was under the control of the opponent, the opponent was *expressly or impliedly notified,* by the *pleadings or otherwise,* that the writing would be needed at the hearing, and *on request at the hearing* the opponent has *failed to produce* the writing [italics added]."

Another convenience-based exception to the original-document rule is found in California Evidence Code section 1509. There, an exception is allowed for circumstances where the papers are "numerous accounts or other writings that cannot be examined in court without great loss of time, and the evidence sought from them is only the general result of the whole." It should be remembered that this exception overcomes only the original-document rule. If the original writings are inadmissible *hearsay* (e.g., business records not sufficiently authenticated), the summary is inadmissible for that reason. Section 11509 also qualifies its exception: The judge in his discretion *may require* that such accounts or other writings be *produced for inspection* by the adverse party."

The final exception gives a desirable alternative to the party whose original business records cannot be left with the court as exhibits for an indefinite time without great inconvenience. He may produce the original writing and make it available for the adverse party's inspection, along with the copy; then he may be permitted to *take the original back and leave only the copy.*[43] The official comment to this exception adds: "Of course, if the original shows erasures or other marks of importance that are not apparent on the copy, the adverse party may place the original in evidence himself."

Several states (California notable among them, as a consequence of the San Francisco earthquake and fire of 1906) have statutes that address the problem of records destroyed by "conflagration or other public calamity." They may provide for secondary evidence of the contents of the destroyed writing to be admitted into evidence[44] or in other cases for special court proceedings to "re-establish" the original through order of the court.[45]

C.   CERTIFICATION OF DOCUMENTS

It has been mentioned that the use of a certified copy of certain documents can overcome some of the problems, both practical and evidentiary, which attend the introduction of documents at trial. The value of obtaining certified copies of any government document that might be used in a trial of a title dispute ought to be apparent from the preceding remarks. It may serve to reiterate some of those remarks together. It will be, of course, a rare occurrence

---

43.   Cal. Evid. Code § 1510.
44.   *See, e.g.,* Cal. Evid. Code § 1601.
45.   *See, e.g.,* Cal. Code Civ. Proc. § 1953.10 *et seq.*

that a surveyor or his attorney possesses the original of a government document which is sought to be introduced into evidence. Consequently, one invariably must consider those evidence problems peculiar to introducing a copy of an original document. Those problems are essentially two, insofar as the use of a certified copy is concerned:

1. *Authentication:* The document must be shown to be in fact what its propounder contends it is. (If a party contends a document is a copy of the original plat of an official government survey, evidence must be introduced to establish that fact.)

2. *The original-document rule:* Before a copy of a document may be received into evidence, the original must be accounted for. (Using the same example as before, it must be shown that the original plat of the survey is on file with the Bureau of Land Management, the National Archives, or some other official repository, or destroyed.)

The value of a certified copy of a government document is that in one inexpensive stroke, these two evidence problems are eliminated. In point of fact, a third evidentiary problem is also overcome by the modern rules pertaining to the certification of documents. The certification is simply a statement by the public employee having custody of the original of the document that the copy is a true copy of the original. This statement, since it is offered for the truth of its assertion, is hearsay (*q.v.*); it is nonetheless permitted to be received for this purpose.[46] (There may be other problems, it should not be forgotten, which would be associated with introducing even the original. The original may not be relevant, or it may constitute hearsay.) A document properly certified as a true and correct copy of the original becomes "self-authenticating," and the original-document rule is satisfied because the original is accounted for.[47] Without certification, these evidence rules would probably require calling to the witness stand the records keeper of the agency having possession of the original document to establish the authenticity of the copy and the whereabouts of the original—a rather expensive and time-consuming process.

What is "certification?" It is a written statement, usually of the custodian of records or his deputy, that the document is a correct copy of the original, which is in his possession. It is preferable to have the certificate acknowledged by a notary public or accompanied by the seal of the agency, although not all jurisdictions have these requirements.[48] Most public agencies have a standard form for certifying documents, which may include colored ribbons or an ornate seal. If the custodian is unsure how to certify a copy of a document or if there are questions about the adequacy of a certificate, the attorney should be consulted.

Frequently one seeks to show the absence of a public record. One may want to show that a township has not been surveyed by the government, for example, or that no patent has issued for a particular tract of land. Proving that something does not exist can obviously be far more difficult than proving that something does, but the law has provided a somewhat

---

46. *See, e.g.,* Cal. Evid. Code § 1601.
47. Cal. Evid. Code §§ 1530, 1531; Fed. R. Evid. 902, 1005.
48. *See* Cal. Evid. Code §§ 1530, 1452, 1453; Fed. R. Evid. 902(1), (2).

painless method, similar to the use of certified copies. A written statement by the custodian of records, "reciting diligent search and failure to find a record," may be introduced to prove the nonexistence of the record.[49] Again, it is preferable though not necessary in all jurisdictions that the statement be under seal or acknowledged by a notary public. A few agencies have a preprinted form for such a document. The Bureau of Land Management calls its form a "44(b) Certificate," so named for Rule 44(b) of the Federal Rules of Civil Procedure, which provides for such a statement.

Last it should be emphasized that any certificate that a document is a true copy, or that no document exists, is not conclusive. It merely raises a presumption,[50] which may be rebutted by evidence that the certificate was forged, that the copy does not comport with the original, or that it is in some other respect deficient.

In addition to government documents, certain private writings may also be proved by means of certified copies, so that the original need not be produced. For the purpose of land cases, the most important such writings are those recorded documents affecting interests in real property. California Evidence Code section 1507 provides, for example:

> A copy of a writing is not made inadmissible by the best evidence rule if the writing has been recorded in the public records and the record or an attested or certified copy thereof is made evidence of the writing by statute.

The reader may have noted the second conditional clause in section 1507: " . . . if the . . . certified copy thereof is made evidence of the writing by statute." Precisely how the use of a certified copy of the recorded copy of a deed meets the authentication requirement is a many-stepped process. For the reader interested in such involutions, the following remarks of the California Assembly Committee deliberating the state's proposed new Evidence Code in 1965 are offered:

> In some instances, however, authentication of a copy will provide the necessary evidence to authenticate the original writing at the same time. For example: If a copy of a recorded deed is offered in evidence, Section 1401 requires that the copy be authenticated—proved to be a copy of the official record. It also requires that the official record be authenticated—proved to be the official record—because the official record is a writing of which secondary evidence of its content is being offered. Finally, Section 1401 requires the original deed itself to be authenticated—proved to have been executed by its purported maker—for it, too, is a writing of which sec-

---

49. Cal. Evid. Code § 1284; *see also* Fed. R. Evid. 803(10); Fed. R. Civ. Proc. 44(b). Some recent cases include *United States v. Alexander*, 48 Fed.3d. 177 (9th Cir. 1995) (FDIC records contained no notation that bank's insurance had been canceled); *United States v. Hutchinson*, 37 Fed.3d 1068 (9th Cir. 1993) (certificate that IRS records reflected no tax return). It may be dangerous to point only to Ninth Circuit decisions, however. More often than any other circuit, it is reversed by the Supreme Court.
50. *See, e.g.,* Cal. Evid. Code § 1530(b).

ondary evidence of its content is being offered. The copy offered in evidence may be authenticated by the attestation or certification of the official custodian of the record as provided by Section 1530. Under Section 1530, the authenticated copy is prima facie evidence of the existence and content of the official record itself. Thus, the authenticated copy supplies the necessary authenticating evidence for the official record. Under Section 1600, the official record is prima facie evidence of the existence and content of the original deed and of its execution by its purported maker; hence, the official record is the requisite authenticating evidence for the original deed. Thus, the duly attested or certified copy of the record meets the requirement of authentication for the copy itself, for the official record, and for the original deed.

Rule 902(4) of the Federal Rules of Evidence also treats as self-authenticating a certified copy of "a document authorized by law to be recorded or filed and actually recorded or filed in a public office. " In turn, Rule 1005 provides that such a certified copy solves the original document rule problem federal courts.

A host of other special provisions for proof of official records have been made from time to time by Congress. The Advisory Committee on the Federal Rules of Civil Procedure drafted a note to Rule 44, containing a useful list of such provisions. It is reprinted here verbatim for the reader's benefit:

NOTES OF ADVISORY COMMITTEE ON RULES

This rule provides a simple and uniform method of proving public records, and entry or lack of entry therein, in all cases including those specifically provided for by statutes of the United States. Such statutes are not superseded, however, and proof may also be made according to their provisions whenever they differ from this rule. Some of those statutes are:

U.S.C., Title 28, former:

§ 661	(Copies of department or corporation records and papers; admissibility; seal)
§ 662	(Same; in office of General Counsel of the Treasury)
§ 663	(Instruments and papers of Comptroller of Currency; admissibility)
§ 664	(Organization certificates of national banks; admissibility)
§ 665	(Transcripts from books of Treasury in suits against delinquents; admissibility)
§ 666	(Same; certificate by Secretary or Assistant Secretary)
§ 670	(Admissibility of copies of statements of demands by Post Office Department)
§ 671	(Admissibility of copies of post office records and statement of accounts)

§ 672   (Admissibility of copies of records in General Land Office)

§ 673   (Admissibility of copies of records, and so forth, of Patent Office)

§ 674   (Copies of foreign letters patent as prima facie evidence)

§ 675   (Copies of specifications and drawings of patents admissible)

§ 676   (Extracts from Journals of Congress admissible when injunction of secrecy removed)

§ 677   (Copies of records in offices of United States consuls admissible)

§ 678   (Books and papers in certain district courts)

§ 679   (Records in clerks' offices, western district of North Carolina)

§ 680   (Records in clerks' offices of former district of California)

§ 681   (Original records lost or destroyed; certified copy admissible)

§ 682   (Same; when certified copy not obtainable)

§ 685   (Same; certified copy of official papers)

§ 687   (Authentication of legislative acts; proof of judicial proceedings of State)

§ 688   (Proofs of records in offices not pertaining to courts)

§ 689   (Copies of foreign records relating to land titles)

§ 695   (Writings and records made in regular course of business; admissibility)

§ 695e   (Foreign documents on record in public offices; certification)

U.S.C., Title 1:

§ 112   (Statutes at large; contents; admissibility in evidence)

§ 113   ("Little and Brown's" edition of laws and treaties competent evidence of Acts of Congress)

§ 204   (Codes and supplements as establishing prima facie the laws of United States and District of Columbia, etc.)

§ 208   (Copies of supplements to Code of Laws of United States and of District of Columbia Code and supplements; conclusive evidence of original)

U.S.C., Title 5:

§ 490   (Records of Department of Interior; authenticated copies as evidence)

U.S.C., Title 6:

§ 7   (Surety Companies as sureties; appointment of agents; service of process)

U.S.C., Title 8:

§ 9a   (Citizenship of children of persons naturalized under certain laws; repatriation of native-born women married to aliens prior to September 22, 1922; copies of proceedings)

§ 1443   (Regulations for execution of naturalization laws; certified copies of papers as evidence)

§ 1443   (Certifications of naturalization records; authorization; admissibility U.S.C., Title 11:

§ 44(d), (e), (f), (g)   (Bankruptcy court proceedings and orders as evidence)

U.S.C., Title 15:

§ 127   (Trade-mark records in Patent Office; copies as evidence)

U.S.C., Title 20:

§ 52   (Smithsonian Institution; evidence of title to site and buildings)

U.S.C., Title 25:

§ 6   (Bureau of Indian Affairs; seal; authenticated and certified documents; evidence)

U.S.C., Title 31:

§ 46   (Laws governing General Accounting Office; copies of books, records, etc., thereof as evidence)

U.S.C., Title 38:

§ 11g   (Seal of Veterans' Administration; authentication of copies of records)

U.S.C., Title 40:

§ 238   (National Archives; seal; reproduction of archives; fee; admissibility in evidence of reproductions)

§ 270c   (Bonds of contractors for public works; right of person furnishing labor or material to copy of bond)

U.S.C., Title 43:

§§ 57–59   (Copies of land surveys, etc., in certain states and districts admissible as evidence)

§ 83   (General Land Office registers and receivers; transcripts of records as evidence

U.S.C., Title 46:

§ 823   (Records of Maritime Commission; copies; publication of reports; evidence)

U.S.C., Title 47:

§ 154(m)   (Federal Communications Commission; copies of reports and decisions as evidence)

§ 412   (Documents filed with Federal Communications Commission as public records; prima facie evidence; confidential records)

U.S.C., Title 49:

§ 14(3)   (Interstate Commerce Commission reports and decisions; printing and distribution of copies)

§ 16(13)   (Copies of schedules, tariffs, etc., filed with Interstate Commerce Commission as evidence)

§ 19a(i)   (Valuation of property of carriers by Interstate Commerce Commission; final published valuations as evidence)

It should be borne in mind that the three points treated in this chapter are by no means the only evidence issues that must be considered when seeking to introduce (or to prevent the introduction) of a document. The relevance issue, discussed in Chapter 1, pertains with respect to all evidence; the document may constitute hearsay; and there may be a question whether the document can be judicially noticed. The hearsay rule is taken up next.

Chapter 3

The Rule Against Hearsay, Or, Perhaps, The Rules Permitting Hearsay

Actual evidence I have none,
But my aunt's charwoman's sister's son
Heard a policeman, on his beat,
Say to a housemaid in Downing Street,
That he had a brother, who had a friend,
Who knew when the war was going to end.

—Reginald Arkell, in *All the Rumors* (1916)

A.   INTRODUCTION

"Ask the man on the street what he knows about the law of evidence. Usually the only doctrine he will be able to mention is the one called by the old English word hearsay."[1] Yet hearsay is perhaps the least understood, the subtlest of the concepts of evidence. The same testimony, the same document, may be hearsay if offered for one purpose and not hearsay if offered for another. As with the other evidence principles treated in this book, the purpose of this chapter is not to provide an exhaustive treatment of the hearsay rule, which has been done ably by many others. Its purpose rather is to provide the reader a nodding acquaintance with the rule—what hearsay is, what it is not, and what exceptions to the rule against hearsay may be of use in preparing for trial.

Centuries of common-law decisions have shaped and refined the rule against hearsay as a legal tool to maximize the probability of ascertaining the truth at trial. As the name suggests,

---

1.   McCormick on Evidence § 223, at 455 (1954).

hearsay is the testimony not of the witness on the stand, but of someone else, speaking (or writing) outside of the courtroom. The "witness" thus has not been placed under oath, and so he is not liable for perjury if his testimony is false, nor is he subject to cross-examination. Moreover, the trier of fact (the jury or if there is no jury, the judge) cannot observe the demeanor of the witness, for whatever light his demeanor sheds on his truthfulness. These points and others have been advanced for centuries as reasons for the rule against hearsay.

B. DEFINITIONS

What is hearsay? The reader is urged to put out of mind any present ideas of what hearsay is and to consider the following variously formulated definitions thoughtfully. Professor McCormick defines it as follows:

> Hearsay evidence is testimony in court or written evidence, of a statement made out of court, such statement being offered . . . to show the truth of matters asserted therein, and thus resting for its value upon the credibility of the out-of-court asserter.[2]

The Federal Rules provide the following:

> "Hearsay" is a statement [either written, oral, or nonverbal conduct, if intended as an assertion], other than one made by the declarant while testifying at the trial or hearing, offered in evidence to prove the truth of the matter asserted.[3]

The California definition is as follows:

> "Hearsay evidence" is evidence of a statement that was made other than by a witness while testifying at the hearing and that is offered to prove the truth of the matter stated.[4]

These definitions show that there are two essential components of hearsay evidence. The first is almost self-evident: It is evidence of a statement made out of court. The witness on the stand, for example, relates what someone told him during a meeting. It should be recalled that the "statement" may be written as well as oral. Thus field notes of a survey when introduced at trial are evidence of a statement made out of court. (In most jurisdictions, nonverbal conduct, if intended to assert something, may also constitute a "statement"; a deaf-mute's sign-language statement is an example.)

It is the second component of the hearsay definition, however, that is elusive: The statement must be offered at the trial or hearing for the purpose of proving the truth of the matter asserted in it. If it is offered for another purpose, it is *not* hearsay Some examples may illustrate.

In a murder trial, suppose the defendant testifies that just before he killed the victim, the

---

2. McCormick on Evidence § 225, at 460 (1954).
3. Fed. R. Evid. 801(c).
4. Cal. Evid. Code § 1200 (a)

victim raised a gun and shouted, "I intend to kill you." If the victim's statement were offered to prove the truth of what was stated—that is that he intended to kill the defendant—it would constitute hearsay because it is a statement made out of court, offered for the truth of the matter asserted. If, however, it were offered to show simply that the victim had in fact said the words—irrespective of their truth—and thus had put defendant in fear of his life (which would help establish self-defense), the evidence would not constitute hearsay.

Suppose a witness testifies that another person told him the defendant was driving 70 miles an hour. Again the statement is one made out of court, but is it hearsay? If it is offered to show that defendant was in fact driving at that speed, it is. But it may be offered for other purposes. If it is offered to show that the one reporting the speed of the defendant (the "declarant") could speak English, it is not hearsay. Similarly, if it is offered to prove that the declarant was conscious at the time, it is again not hearsay.

In a California case,[5] evidence of telephone conversations was held admissible to show that the defendant was operating a book-making business. The court reasoned that the conversations did not constitute hearsay because they were not offered to prove the truth of what was said, but simply to show that the conversations took place; because the conversations discussed the placing of bets, they tended to prove the existence of the gambling business.

If a map is offered to show the truth of what is depicted on it, as for example the location of monuments, it is hearsay because it is a "statement" made out of court and is offered to prove the contents of the statement. If the question in the case is simply whether the map exists—regardless of what it depicts—it does not constitute hearsay.

The critical factor in these examples is whether the evidence of the out-of-court statement is offered to prove the truth of what was said or merely to show the making of the statement. This is why an out-of-court statement may be objectionable when offered to show one thing, but not when offered to show another; the trier of fact may or may not need to consider the truthfulness of the out-of-court declarant. If the jury must decide whether the car was being driven at 70 miles an hour, to use one of the examples, it needs to know whether the out-of-court declarant was telling the truth. He was neither placed under oath nor subjected to cross-examination when he made the statement. Thus two of our legal system's most revered guarantors of truth are lacking. If the jury on the other hand is not concerned with the speed of the car, but only with whether the declarant was conscious, it has no need to know whether the declarant was speaking the truth about the speed of the car. It need only know whether the witness on the stand is telling the truth when he testifies that the declarant spoke; and the witness on the stand has taken the oath and may be cross-examined.

C.   MULTIPLE HEARSAY

Hearsay within hearsay is common. A newspaper account reads, "According to the police report, witnesses told the investigating officers that the robbery occurred just before midnight Wednesday." If the witnesses mentioned in the police report were called to the witness stand to testify when the robbery occurred, and their testimonies were offered to prove the time of the robbery, their testimonies would of course not be hearsay. If the officers testified, for the same purpose, as to what those eyewitnesses told them, their testimony would be

---

5. *People v. Radley*, 68 Cal.App.2d 607, 609 (1945).

hearsay. The police report, if offered for the same purpose, would constitute hearsay upon hearsay. And the newspaper account is triple hearsay.

Is multiple hearsay ever admissible? It is, so long as an exception to the hearsay rule applies to each item of hearsay within the statement. The business-records exception, for example, would apply to the police report. But there is probably no exception that would admit the hearsay statements of the investigating officers as to what those eyewitnesses told them; nor is there a likely exception that would admit the newspaper account. Thus, neither the police report nor the newspaper account would be admissible. In other situations, an exception may apply at each level of hearsay and permit the admission of multiple hearsay. Suppose a land surveyor writes a memorandum to himself that, in the course of surveying a parcel of land, the landowner told him that his property was bounded by a certain fence and did not extend to the riverbank. The statement of the landowner is hearsay if offered to prove the location of his boundary, and the surveyor's memorandum is hearsay upon hearsay. But in litigation concerning the location of the boundary, the memorandum might be admitted because the landowner's statement is an "admission," and the memorandum recording the statement might meet the requirements of past recollection recorded, each of which is an exception to the hearsay rule.[6]

D.   RATIONALE

As Professor McCormick has long ago written, "[t]he common law system of proof is exacting in its insistence upon the most reliable sources of information,"[7] and the rule against hearsay is a product of that insistence. Traditionally, three reasons are adduced as establishing the unreliability of hearsay. First, the declarant had not sworn to tell the truth when he made the out-of-court statement. While it may be questioned today whether the oath alone impels the average witness to be truthful when he would otherwise lie, the penalties for perjury may be a real inducement.

The second reason is that the trier of fact cannot observe the declarant's demeanor when he made the statement. Was the declarant calm and straightforward or was he nervously wringing his hands? The answer to such a question, obviously, may greatly influence one's assessment of the truth of the witness's statement.

Finally, the declarant is not subject to cross-examination, "the greatest legal engine ever invented for the discovery of truth."[8] Suppose a witness on the stand testifies, "My father told me the sycamore tree marked the northeast corner of the farm." The father is in effect testifying, and the opposing attorney cannot cross-examine him to determine whether he had a sound basis for making the statement, whether he had a motive to lie, and so forth.[9]

---

6. *See, generally,* Fed. R. Evid. 805.
7. McCormick on Evidence § 10, at 19 (1954).
8. 5 Wigmore on Evidence § 1367, at 32 (rev. ed. 1974).
9. For a general discussion of the reasons for the hearsay rule, *see* McCormick on Evidence § 224, at 457–459 (1954). The practice of placing the witness under oath has an interesting and sometimes humorous history. *See* P. Devlin, Trial by Jury 5–7 (1956); 1 W. Holdsworth, A History of English Law 305–308 (7th rev. ed. 1956).

E.   THE MANIFOLD EXCEPTIONS TO THE HEARSAY RULE

The general rule, of course, is that hearsay evidence is not admissible.[10] But many exceptions have been made. The Federal Rules of Evidence, for example, provide no fewer than twenty-some exceptions. The last of which, which is now given a whole rule of its own, is that the hearsay rule does not exclude:

> a statement not specifically covered by any of the foregoing exceptions but having equivalent circumstantial guarantees of trustworthiness, if the court determines that (A) the statement is offered as evidence of a material fact; (B) the statement is more probative on the point for which it is offered than any other evidence which the proponent can procure through reasonable efforts; and (C) the general purposes of these rules and the interests of justice will best be served by admission of the statement into evidence. . . .[11]

The exceptions to the hearsay rule are fairly uniform among the states and in the federal system. (That is easy to assert, since 39 states have adopted the Federal Rules.) Some exceptions deal specifically with criminal matters, and others would rarely be encountered by the reader. Accordingly, the list below is not intended as exhaustive, but emphasizes those that may arise in land and natural-resource cases. Also, many of the exceptions do not operate unless certain rather technical matters are first shown. These "foundational" matters will be discussed only sparingly.

ADMISSIONS OF A PARTY

Evidence of a statement made out of court by a party to a lawsuit may be introduced, for the purpose of proving the truth of the statement, by an opposing party.[12] It may not be used, however, by the party who made the statement. Suppose the plaintiff in an adverse-possession case once told a neighbor that his occupation had been intermittent and not continuous. The defendant may call the neighbor to the stand and have him testify to what the plaintiff told him, and it is admissible for the truth of the matter asserted; that is, it may be considered as evidence that the plaintiff's possession had not been continuous. The reason for this exception is apparent. The plaintiff is a party to the action and, thus, is available to explain or deny the statement. Professor McCormick writes, "This notion that it does not lie in the opponent's mouth to question the trustworthiness of his own declarations is an expression of feeling rather than logic but it is an emotion so universal that it may stand for a reason. The feeling that one is entitled to use the opponent's words is heightened by our contentious or adversary system of litigation."[13]

Two types of "admissions" should be distinguished. A judicial admission is a statement contained in a pleading or in certain other documents filed in the lawsuit, such as a

---

10.   Fed. R. Evid. 802; Cal. Evid. Code § 1200 (b).
11.   Fed. R. Evid. 803 (24) and 805 (5), now Rule 807.
12.   Cal. Evid. Code § 1220. The federal rules interestingly treat admissions as nonhearsay, as opposed to treating them as exceptions to the rule. Fed. R. Evid. 801 (d)(2).
13.   McCormick on Evidence § 239, at 503 (1954).

stipulation. Suppose in response to a complaint alleging adverse possession the defendant files an answer admitting that plaintiff has paid all taxes assessed on the disputed property, but denies that plaintiff has met the other requirements for title by adverse possession. The statement respecting the payment of taxes is a judicial admission, but it is not the kind of admission that constitutes an exception to the hearsay rule. A judicial admission removes the factual question—in this case, whether plaintiff had in fact paid all assessed taxes— from the lawsuit. It is deemed established, and no evidence on the matter can be received at trial, because it would be irrelevant. On the other hand, suppose that the defendant files an answer denying all allegations of the complaint and that the plaintiff calls to the stand a witness who testifies that he heard the defendant say (out of court) that the plaintiff had paid all taxes assessed on the subject property. This is an "evidential admission," in Professor McCormick's terminology; it is hearsay, but because it is the admission of an opposing party it is admissible as an exception to the hearsay rule. Unlike a judicial admission, however, the evidential admission is not conclusive. The defendant would still be able to show (by the use of the tax collector's records, for example) that his admission was incorrect and that plaintiff had not in fact paid all of the taxes. Such evidence would be inadmissible had the defendant made a judicial admission (unless the court allows him to amend his pleading to remove the admission).

Hearsay evidence is sometimes admissible when it consists of the out-of-court admission, not of the opposing party, but of his agent or someone else identified with the party. Generally, the rules require that the agent or representative be specifically authorized by the party to speak on the subject or that the statement be made directly within the scope of the agent's duties for the party.[14]

An admission may also occur when a party has manifested his belief in a statement made by another person.[15] Suppose a party to a boundary lawsuit is shown to have commissioned a survey of his property, and later moved his fences to correspond to the lines as surveyed. At trial he contends for a different placement of the boundary. His conduct in assenting to the lines as surveyed may be shown as an admission of the correctness of those lines. (Recall that conduct, as well as verbal and written statements, may constitute hearsay.) This is sometimes called an "adoptive admission."

One kind of "vicarious admission" particularly worthy of note in land cases is the statement of a predecessor in title.[16] This situation arises when a party is required to establish his chain of title. In order to establish his title, the party must show that the title at some time in the past was vested in X, an admission of X made during the time the party claims that X held the title is admissible against the party. If the statement was made before or after X was the alleged title-holder, the statement is not admissible. Suppose that the plaintiff in a quiet-title action has introduced documents purportedly constituting a chain of title showing that between 1929 and 1941 the fee was vested in McDermott. Letters of McDermott to a friend written during this period are found in which McDermott complains that he has not been able to perfect his title, that residual interests from some prior disposition of the property are outstanding. McDermott's letters may be introduced into evidence by the defendant.

---

14. *See* Fed. R. Evid. 801(d)(2)(C) & (D); Cal. Evid. Code §§ 1222–1223.
15. Fed. R. Evid. 801(d)(2)(B); Cal. Evid. Code § 1221.
16. Cal. Evid. Code § 1225.

DECLARATIONS AGAINST INTEREST

The hearsay exception for admissions is so frequently confused with the exception for declarations against interest that the erroneous expression "admissions against interest" is often uttered. While the same hearsay statement may constitute both an admission and a declaration against interest, the two have different elements. Essentially a declaration against interest must be one that, when made, was antithetical to the declarant's pecuniary or proprietary interest, and the declarant must be unavailable to testify at trial. An admission, on the other hand, need not have been against the declarant's interest at the time it was made (although it usually is). Moreover, the exception for admissions applies only to statements by parties to the lawsuit, whereas the exception for declarations against interest applies to any declarant.

The rationale for allowing hearsay evidence of declarations against interest is aptly stated in the California statute codifying the exception:

> Evidence of a statement by a declarant having sufficient knowl-
> edge of the subject is not made inadmissible by the hearsay rule if the
> declarant is unavailable as a witness and the statement, when made,
> was so far contrary to the declarant's pecuniary or proprietary inter-
> est, or so far subjected him to the risk of civil or criminal liability, or
> so far tended to render invalid a claim by him against another, or cre-
> ated such a risk of making him an object of hatred, ridicule, or social
> disgrace in the community, that a reasonable man in his position
> would not have made the statement unless he believed it to be true.[17]

Examples in the real-property field of declarations against interest are common. Statements by one in possession of real property that he is not the true owner, that he has fenced beyond his true boundaries, or that the property had been purchased with trust funds have been traditionally considered declarations against interest.[18] Examples in other areas of law are admitting liability for an automobile accident, acknowledging a debt, or acknowledging a breach of contract. The belief that such statements would not be made unless true overcomes the infirmities of the lack of oath and opportunity for cross-examination.

In all such cases, however, the declarant must be unavailable to testify at trial. What "unavailable" means varies from jurisdiction to jurisdiction. If he is dead, however, beyond the subpoena power of the court, or physically incapacitated, all courts would consider him "unavailable."

BUSINESS RECORDS

Certain entries in business records are considered admissible evidence in most jurisdictions, notwithstanding that they are hearsay.[19] Some such entries, of course, might come in under other exceptions to the hearsay rule. If the entries were made by a party opponent, they

---

17. Cal. Evid. Code § 1230; *see also* Fed. R. Evid. 804(b)(3).
18. *Lamar v. Pearre*, 90 Ga. 377, 17 S.E. 92 (1892); *Smith v. Moore*, 142 N.C. 277, 55 S.E. 275 (1906); *Carr v. Bizzell*, 192 N.C. 212, 134 S.E. 462 (1926); *Barlow v. Greer*, 222 S.W. 301 (Tex. Civ. App. 1920).
19. Cal. Evid. Code § 1271; Fed. R. Evid. 803(6).

may be admissible as admissions; perhaps the entries, whether made by a party or not, were declarations against interest at the time they were made. Because of the general reliability of records kept in the regular order of business, however, a specific exception to the hearsay rule for such records has been made. The rule no doubt originated with the admission of a shopkeeper's books to show the making of a debt; without his books, the shopkeeper could not testify as to the creation of the debt except by independent recollection.[20]

Today, evidence of entries in business records is admissible so long as certain matters are shown. (These requirements may vary somewhat from jurisdiction to jurisdiction.) First, the writing must have been made in the regular course of business. The accounting records of a business are the most obvious example, and perhaps the earliest as well. A doctor's records that disclose the findings of a physical examination of his patient are generally admissible, because it is customary to make such records following examinations. The United States Supreme Court has ruled, however, that a railroad engineer's report of an accident was properly excluded from evidence, reasoning that the report was made in contemplation of litigation and not "for the systematic conduct of the enterprise as a railroad business."[21] Because railroads and virtually every other transportation enterprise have regular accident-reporting procedures, the Court's concern was probably not so much that the record was not made in the regular course of the business, but rather that it was inherently self-serving. Addressing this concern, many jurisdictions have imposed a separate requirement that the record have the earmarks of reliability. This requirement is discussed below.

Second, the writing must have been made at or near the time of the act or condition that is recorded. A doctor's report detailing his findings upon examining a patient is typically made either while he is examining the patient or immediately afterward. Likewise, the field notes of a survey are made as the surveyor is actually running the lines in the field. Entries in field tablets that were not made until some time after the survey was run—corrections in mathematics, for example, or observations of physical features—may be inadmissible under this exception.

Third, the custodian of the document or some other qualified witness must testify as to the identity of the document and its mode of preparation. The best-qualified witness, of course, is the one who actually made the entry, but it is not always possible, nor necessary, to call this person to the stand. If a physician is deceased, his office nurse could identify a report as having been made by the physician and confirm that he invariably made such reports while or immediately after examining a patient. The custodian of records of a district office of the Bureau of Land Management can testify how its staff posts entries on master title plats, laying the "foundation" for admissibility of entries from those records. (There are other avenues available for introducing such entries, as is true for many of the remarks made here on how an evidentiary problem may be overcome.)

A fourth requirement, alluded to earlier, is that the records bear the earmarks of reliability. The federal rules, for example, specify that business records are inadmissible if "the source of information or the method or circumstances of preparation indicate lack of trustworthiness."[22] It is difficult to generalize on circumstances that indicate a lack of trustworthi-

---

20. 5 Wigmore on Evidence § 1518 (rev. ed. 1974); *Radtke v. Taylor,* 105 Or. 559, 210 P. 863 (1922).

21. *Palmer v. Hoffman,* 318 U.S. 109, 114 (1943).

22. Fed. R. Evid. 803(6).

ness, but self-serving declarations certainly fall within this category.[23] Erasures or other alterations of records, or mutilations such as torn-out pages, also cast suspicion on the reliability of the records. If the erasure, alteration, or mutilation can be explained to the satisfaction of the court, it may nonetheless be received into evidence, but these conditions make more difficult the task of introducing such records.[24]

In this respect, two well-known authors have some sound suggestions for the making of survey field notes, which a land surveyor should heed:

> When a survey is performed for the purpose of gathering data, then the field notes become the record of the survey. If the notes have been carelessly recorded and documented, falsified, lost, or made grossly incorrect in any way, the survey or a portion of the survey is rendered useless. Defective notes result in tremendous waste of both time and money. Furthermore, it will become obvious that, no matter how carefully the field measurements are made, the survey as a whole may be useless if some of those measurements are not recorded or if the meaning of any record is ambiguous.
>
> The keeping of neat, accurate, complete field notes is one of the most exacting tasks. Although several systems of note-keeping are in general use, certain principles apply to all.
>
> *   *   *
>
> A hard pencil—4H or harder—should be used to prevent smearing. *No erasures should be made,* because such notes will be under suspicion of having been altered.
>
> Clear, plain figures should be used, and the notes should be lettered rather than written. *The record should be made in the field book at the time the work is done,* and not on scraps of paper to be copied into the book later. Copied notes are not original notes, and there are too many chances for making mistakes in copying or of losing some of the scraps.[25]

Thus if it is sought to be proved that Joseph Wilkes visited a title abstractor on a certain day, then the abstractor's records of appointments made and kept (assuming those records meet the requirements that have been discussed) are admissible to establish the fact. Suppose, however, that it is sought to be proved that Wilkes did not visit his abstractor on a certain day. A corollary to the business-records exception allows introduction of the abstractor's

---

23. *See Cummins v. Pennsylvania Fire Ins. Co.,* 153 Iowa 579, 134 N.W. 79 (1912), in which the court held that memoranda entered by an agent of the insurance company, since deceased, in his "policy register," were self-serving statements and hence inadmissible.

24. *See* Annot., 142 A.L.R. 1406.

25. Moffitt and Bouchard, *Surveying* (6th ed., 1975). [Emphasis added.]

appointment book to show the *absence* of an entry that would have recorded his visit.[26] Again, it must be shown that in the normal course of business a visit by Wilkes would have been recorded, and the records must have the indicia of trustworthiness. If Wilkes claims he arrived at the abstractor's office at 3:00 p.m., and the appointment book has an erasure for that time, it is likely the book will be ruled inadmissible.

While hearsay is generally thought to be inferior evidence, business records, being contemporaneous accounts of events, in many instances are accorded special status. Thus federal law provides that a court in admiralty may dismiss charges against a seaman when an account of the alleged offense was not entered in the ship's official log.[27]

For the business records exception to apply, it might be added, it is not necessary that the declarant—the one who made the entry—be unavailable to testify.

CERTAIN OFFICIAL DOCUMENTS

An exception to the hearsay rule, akin to the exception for business records, has been made for certain official documents.[28] Such matters as court records, records of acts of the legislature, observations recorded by the National Weather Service, and land-office records fall within this exception.

The requirements for the exception are similar to those for the business records exception. The public employee who made the writing must have done so within the scope of his duty. A land office clerk has the duty to record all applications for patents, for example, but it is beyond the scope of his duty to record that the applicant has a suspicious appearance.[29] Second, the writing must have been made at or near the time of the event recorded. Thus it is helpful to have not only the date of the event recorded, but also the date the entry was made. Testimony that the land office clerks habitually recorded applications on the day received would suffice, however, if the date the entry was made does not appear. Third, the circumstances must indicate the trustworthiness of the entry in question. Again, erasures, alterations, and mutilations may obstruct the introduction of an official record into evidence.

The reasons for the exception admitting government documents parallel those given for the business-records exception. If it is customary to make a record of the kind of occurrence in question, if the record was made near in time to the occurrence, and if there are no circumstances to indicate a lack of trustworthiness, the record is no doubt better evidence of what occurred than what the writer could testify to from memory. In the case of public records, there is also the consideration that summoning the writer to the witness stand would unnecessarily interrupt public business.[30]

There is also an exception to the hearsay rule which admits evidence of the *absence* of a public record of an event that was required to be recorded. The evidence is admissible to prove the nonoccurrence of an event (the records of the National Weather Service contain no mention of a hurricane on the date in question) or the nonexistence of a document (the

---

26.   Cal. Evid. Code § 1272; Fed. R. Evid. 803(7).
27.   46 U.S.C. § 11502; *The Amazon*, 144 F. 153 (N.D. Wash. 1906).
28.   *See, e.g.,* Cal. Evid. Code §§ 1280–1281; Fed. R. Evid. 803(8) & (9).
29.   The presumption that "official duty has been regularly performed" also operates in this context of proving official acts, or the lack of an official act. *See, e.g.,* Cal. Evid. Code § 664.
30.   McCormick on Evidence § 291, at 615 (1954).

records of the land office do not contain an application by Jones for a patent). Strictly speaking, this rule perhaps ought not to be characterized as an exception to the hearsay rule, because the *absence* of a record is not a "statement." On the other hand, the entirety of the official records is a "statement" of events that have occurred and, together with its silence on the event in question, is an implied "statement" that the event did not occur. Convenience in any event dictates that it be treated together with the exception for official writings.

The evidence showing the absence of a public record may consist either of the testimony of the custodian of records or of a certified statement by him that the record does not exist. (*See* Certification of Documents, Section C in Chapter 2.) The testimony or statement should recite that a diligent search has been made to locate the record in question and it has not been found. Certain agencies have standard forms for this purpose. The Bureau of Land Management uses a form "44(b)" to provide evidence that land has not been patented, for example.

PAST RECOLLECTION RECORDED

Another exception to the hearsay rule, which in many respects resembles the exception for business records, addresses what is generally called "past recollection recorded."[31] According to this exception, evidence of what is contained in a memorandum or other writing may be admitted into evidence if certain elements are shown. First the writing must have been made by the witness himself, or by someone under his direction, or by some other person for the purpose of recording the witness's statement at the time it was made (as, for example, a shorthand reporter taking the statement of a witness after a crime).

Second, the writing must have been made when the fact recorded in the writing actually occurred or was fresh in the witness's memory. Generally, this may be shown simply by having the witness testify that the statement he made was a true statement of the fact recorded.

Third, the witness must presently have insufficient recollection to testify fully and accurately to the facts recorded in the writing.

With few exceptions, when these "foundational" requirements have been met, courts do not admit the writing itself into evidence, but rather allow it to be read into the record (as though the witness were testifying to the matters recorded). If, however, the writing itself is offered by an adverse party, generally it may be received into evidence.

From the foregoing discussion, it should be apparent that a surveyor's properly recorded field notes are generally admissible into evidence as past recollection recorded, and as business records as well. So also are an appraiser's notes, or those of wetland scientists.

REPUTATION IN GENERAL

Evidence of community reputation on various matters is admissible by virtue of exceptions to the hearsay rule. The practice of receiving evidence of community reputation in common law courts antedates even the development of a hearsay rule, which had emerged only by the end of the seventeenth century. At that time it was a traditional right of a jury to resort to common reputation within its community. As Wigmore has noted, it would have been "unnatural and improbable" to have expected a jury to disregard what they knew by community reputation.[32]

---

31.   Cal. Evid. Code § 1237; Fed. R. Evid. 803(5).
32.   5 Wigmore on Evidence § 1580, at 544 (rev. ed., 1974).

Today, evidence of community reputation is admissible under exceptions to the hearsay rule when the reputation concerns the following:

1. Boundaries of land or customary rights in land,
2. Events of general history in the community,
3. Personal character, or
4. Marriage, death, and other facts of family history.

As Wigmore has also observed, while each of the reputation exceptions to the hearsay rule has emerged from a separate line of precedent, there are two common principles supporting the exceptions: "[t]he principle of necessity and the principle of a circumstantial probability of trustworthiness."[33] The necessity arises from the lack of other satisfactory evidence of the fact in question. Land boundaries and customs affecting land, for example, often can be shown only by community reputation. The circumstantial probability of trustworthiness arises from the perception that a reputation is substantial and likely emerged from statements of persons with firsthand knowledge and from thorough discussion in the community. Even though exaggerated and embellished legends often result from the same process (the stories of the Cyclops and the Sirens in Homer's *Odyssey* were handed down over generations by Greek oral tradition), exceptions for reputation evidence have survived to the present date.

REPUTATION CONCERNING LAND BOUNDARIES AND LAND CUSTOMS

The California rule respecting land boundaries and customs is typical of those found in most jurisdictions today:

> Evidence of reputation in a community is not made inadmissible
> by the hearsay rule if the reputation concerns boundaries of, or cus
> toms affecting, land in the community and the reputation arose before
> controversy.[34]

Formerly it was a requirement for the admissibility of reputation evidence concerning land boundaries or customs that the reputation must have been "ancient"—that is, that it must have arisen during past generations.[35] Today, however, the generally accepted principle is simply that the reputation must have arisen in the community before the present controversy arose; there is no requirement that the reputation be "ancient."

The general principle for the rule admitting reputation evidence is this:

> [T]he fact sought to be proved . . . . [being][o]f too ancient a
> date to be proved by eye-witnesses, and not of a character to be made
> a matter of public record, unless it could be proved by tradition, there

---

33. *Id.*, § 1581, at 545.
34. Cal. Evid. Code § 1322; *see also* Fed. R. Evid. 803(20).
35. *See, e.g., Shutte v. Thompson,* 82 U.S. (15 Wall.) 151 (1873); *Dawson v. Town of Orange,* 78 Conn. 96, 61 A 101 (1905); *Mechanics' Bank & Trust Co. v. Whilden,* 175 N.C. 52, 94 S.E. 723 (1917).

would seem to be no mode in which it could be established. It is a universal rule, founded in necessity, that the best evidence of which the nature of the case admits is always receivable.[36]

This principle has special application in boundary cases:

> It must be obvious, that when the country becomes cleared and in a state of improvement, it is oftentimes difficult to trace the lines of a survey made in early times.[37]
>
> Questions of boundary, after the lapse of many years, become of necessity questions of hearsay and reputation. For boundaries are artificial, arbitrary, and often perishable; and when a generation or two have passed away, they cannot be established by the testimony of eye-witnesses.[38]

It should be emphasized that this exception to the hearsay rule does not concern individual statements respecting land boundaries (which statements are treated separately below), but rather the prevalent belief of the community reputation. The reputation may be proved in several ways. Individual witnesses may testify that, according to reputation in the community, a certain ditch or road is the boundary of the tract in question. In addition, old maps and old surveys,[39] if they are shown to have been consulted by the community in its dealings with the land, may be admitted as evidence of reputation in the community concerning boundaries. In the same manner, old deeds and leases may be admissible as evidence of community reputation.[40] Books concerning matters of local history may also recite facts that, if shown to reflect community reputation, would be admissible:

> Historical facts, of general and public notoriety, may indeed be proved by reputation; and that reputation may be established by historical works of known character and accuracy. But evidence of this sort is confined . . . and where, from the nature of the transactions, or the remoteness of the period, or the public and general reception of the facts, a just foundation is laid for general confidence.[41]

While boundaries may be proved by reputation evidence, *title* on the other hand may not be so proved.[42]

---

36. *McKinnon v. Bliss,* 21 N.Y. (7 Smith) 206, 218 (1860).
37. *Montgomery v. Dickey,* 2 Yeates 212 (Pa. 1797).
38. *Harriman v. Brown,* 35 Va. (8 Leigh) 697, 707 (1837).
39. *Taylor v. McConigle,* 120 Cal. 123, 52 P. 159 (1898); *Seaway Co. v. Attorney General,* 375 S.W.2d 923 (Tex.Civ.App. 1964); *Adams v. Stanyan,* 24 N.H. 405 (1852).
40. *Sasser v. Herring,* 14 N.C. (3 Dev.L.) 340, 342 (1832); *Weld v. Brooks,* 152 Mass. 297, 25 N.E. 719 (1890).
41. *Morris v. Lessee,* 32 U.S. (7 Pet.) 554, 558–559 (1833).
42. *Crippen v. State,* 80 S.W. 372 (Tex. Crim. 1904); *School District of Donegal Township v. Crosby,* 112 A.2d 645 (Pa. Super. 1955); *Henry v. Brown,* 39 So. 325 (Ala. 1905); *Howland v. Crocker,* 89 Mass. (7 Allen) 153 (1863).

In addition to reputation concerning boundaries, the hearsay exception extends to evidence of reputation concerning a myriad of customs affecting land. The origins of the rule were succinctly stated by one federal court in 1895:

> The exception, as it originated in the English courts, was confined to such boundaries as were matters of public concern, and was part of a larger exception to the rule. On questions respecting the existence of manors; manorial customs; customs of mining in particular districts; a parochial modus; a boundary between counties, parishes, or manors; the limits of a town; a right of common; a prescriptive liability to repair bridges; the jurisdiction of certain courts,—matters in which the public is concerned, as having a community of interest, from residing in one neighborhood, or being entitled to the same privileges, or subject to the same liabilities,—common reputation and declarations of deceased persons are received, if made, ante litem motam, by persons in a position to be properly cognizant of the facts.[43]

Evidence of customs affecting land is relevant in several contexts of interest in land cases. First, in cases of prescriptive easements or "implied dedication," it may be relevant to establish that it was "customary" for the public to use a certain road, path, or beach. In California, for example, evidence that the public used a certain parcel of land as if it were public land, and was under the belief that it was, may give rise to a public prescriptive easement, sometimes called an "implied dedication."[44] Thus evidence that the parcel in question is reputed to be public land or that the public was reputed to be in the custom of using the parcel—or that it was not—would be *relevant* evidence. And the *hearsay* objection is met by the exception allowing evidence of such reputation. In an early decision of the United States Supreme Court, the waterfront of the City of New Orleans was held to be dedicated to a public use by virtue of evidence in the nature of reputation:

> The original dedication is proved by the maps in evidence, and by a public use of more than a century. These facts are conclusive. . . . No case of dedication to public use has been investigated by this court, where the right has been so clearly established.[45]

Second, an ancient common law doctrine of custom, whereby rights in land arise by virtue of custom, has been resurrected in recent American land decisions. Its application requires proof which in many instances can be supplied only by evidence of reputation or

---

43. *Robinson v. Dewhurst,* 68 Fed. 336, 337 (4th Cir. 1895).
44. *Gion v. City of Santa Cruz,* 2 Cal.3d 29, 39, 44 (1970); *County of Los Angeles v. Berk,* 26 Cal.3d 201 (1980), *cert. denied,* 449 U.S. 836 (1980); *City of Los Angeles v. Venice Peninsula Properties,* 205 Cal.App.3d 1522 (1988); *City of Long Beach v. Daugherty,* 75 Cal.App.3d 972 (1977), *cert. denied,* 439 U.S. 823 (1978); *County of Orange v. Chandler–Sherman Corp.,* 54 Cal.App.3d 561 (1975); *Richmond Ramblers Motorcycle Club v. Western Title Guaranty,* 47 Cal.App.3d 747 (1975).
45. *New Orleans v. United States,* 35 U.S. (10 Pet.) 662, 718 (1836).

custom. In a broad-reaching decision, the Supreme Court of Oregon in 1969 used the doctrine of custom to establish public rights in all of the dry-sand beaches on the Pacific coast of that state:

> Because many elements of prescription are present in this case, the state has relied upon the doctrine in support of the decree below. We believe, however, that there is a better legal basis for affirming the decree. The most cogent basis for the decision in this case is the English doctrine of custom. Strictly construed, prescription applies only to the specific tract of land before the court, and doubtful prescription cases could fill the courts for years with tract-by-tract litigation. An established custom, on the other hand, can be proven with reference to a larger region. Ocean-front lands from the northern to the southern border of the state ought to be treated uniformly.
>
> The other reason which commends the doctrine of custom over that of prescription as the principal basis for the decision in this case is the unique nature of the lands in question. This case deals solely with the dry-sand area along the Pacific shore, and this land has been used by the public as public recreational land according to an unbroken custom running back in time as long as the land has been inhabited.[46]

The elements of the doctrine of custom, a principle of substantive law, (evidence is "procedural" law), as applied by the Oregon Supreme Court, were taken from the eighteenth-century writings of Sir William Blackstone. Among the requirements are that the custom must have been used so long that "the memory of man runneth not to the contrary." Furthermore, the custom must have been uninterrupted, it must have been peaceable, and acquiesced in, and it must be reasonable. "Customs ought to be *certain*. A custom that lands shall descend to the most worthy of the owner's blood is void; for how shall this worth be determined?" In addition, the custom when established must be compulsory, and it cannot be repugnant to another.[47]

Other recent cases have applied the common law doctrine of custom in land cases.[48] In such cases where rights in land are asserted to exist by virtue of custom, the reputation of the custom may be the only evidence that can establish it, making this exception to hearsay a quite valuable rule.

Customary rights can arise in a number of contexts other than public rights in land. A pueblo's rights in surface waters during the Spanish and Mexican reigns in California were thought to have arisen by virtue of custom.[49] The doctrine of appropriative water rights emerged in California as an exception to the common-law principle of riparian water rights by virtue

---

46.   *State ex rel. Thornton v. Hay,* 462 P.2d 671, 676(Or. 1969).
47.   Blackstone, *1 Commentaries* (Cooley, ed., 1899) 66–72 (1969).
48.   *See United States v. St. Thomas Beach Resort, Inc.,* 386 F.Supp. 769, 772–773 (D. Virgin Islands, 1974); *In re Ashford,* 440 P.2d 76 (Hawaii 1968); *State Highway Commission v. Fultz,* 491 P.2d 1171 (Or. 1971); *City of Daytona Beach v. Tona Rama, Inc.,* 294 So.2d 73, 81 (Fla. 1974) (Boyd, J. dissenting).
49.   *See City of Los Angeles v. City of San Fernando,* 14 Cal.3d 199, 240–241 (1975).

of the customs of miners during the gold rush.[50] Again, early judicial decisions in California announced the public right to mine in public lands, a right asserted to have derived from custom.[51]

As an aside, it might be noted that surveying principles become matters of law by virtue of the customs of surveyors in the community, notwithstanding that new developments in technology may prove more precise than the traditional methods. *Dolphin Lane Assoc. v. Town of Southampton,* 372 N.Y.S.2d 52 (1975), a decision of the highest court of the State of New York, is an example. That case concerned the location of the ordinary high-water mark—the boundary between property owned by private parties and tidelands owned by the Town of Southampton. The town produced impressive scientific evidence to show that the ordinary high-water mark could be located with great precision by observing the species of vegetation found within the salt marsh. The court, however, observed that employing such a method produced a high-water mark that deviated from the one that would result by employing the customary methods of surveyors in the community and held:

> There was uncontroverted testimony here that it was the long-standing practice of surveyors in the Town of Southampton to locate the shore-line boundaries by reference to the line of vegetation [i.e., the seaward edge of vegetation]. To give effect to such uniform practice is not, as the town contends, to delegate arbitrary powers to surveyors to determine property lines; rather it is the obverse, namely, to recognize that property lines are fixed by reference to long-time surveying practice.[52]

STATEMENTS OF DECEASED PERSONS

To be distinguished from the hearsay exception for community reputation concerning boundaries and land customs is a distinct exception for certain individual statements respecting boundaries. While the two hearsay exceptions have common historical roots in many jurisdictions, today the two are usually treated as discrete exceptions to the general rule forbidding the introduction of hearsay evidence.

The general requirements for admitting such declarations are, first, that the declarant is dead and, second, that the circumstances indicate no reason for the declarant to have misrepresented the truth.

The land surveyor will appreciate the fact that necessity was the mother of this exception to the hearsay rule:

> In many of the states, and especially in this state, the territory within their limits was first divided into townships, and these were soon after subdivided into small and distributed between the several proprietors. Almost the only evidence upon the land, to indicate the

---

50. *See United States v. Gerlach Live Stock Co.,* 339 U.S. 725, 746 (1950); *Irwin v. Phillips,* 5 Cal. 140, 146–147 (1855).
51. *Biddle Boggs v. Merced Mining Co.,* 14 Cal. 279, 374–379 (1859).
52. *Dolphin Lane Assoc. v. Town of Southampton,* 372 N.Y.S.2d 52, 54 (1975).

location of the lines either of the townships, or of the division between the proprietors, was marks upon the trees standing thereon, and
these evidences, from lapse of time, accidental causes, and the cutting
off the timber, are almost entirely obliterated. . . . If it be said that it
must be, by the testimony of witnesses who have personal knowledge
of their original location, they cannot be proved at all, as in the great
majority of cases, all such persons are now dead.[53]

From this line of reasoning evolved the requirement that the declarant be shown to be dead.

Where boundary is the subject, what has been said by a deceased person is received as evidence. It forms an exception to the
general rule.[54]

We have in questions of boundary, given to the single declarations of a deceased individual, as to a line or corner, the weight of
common reputation. . . . Whether this is within the spirit and reason
of the rule, it is now too late to inquire.[55]

Certain states do not require that the declarant be deceased, but merely that he be unavailable to testify as a witness. The rule in California is as follows:

Evidence of a statement concerning the boundary of land is not
made inadmissible by the hearsay rule if the declarant is unavailable
as a witness and had sufficient knowledge of the subject. . . .[56]

In *Smith v. Glenn,* 129 Cal. 519, 62 P. 180 (1900), for example, the California Supreme Court
admitted the declaration of an owner made while he was in possession of the land and in the
act of surveying it.

Professor Wigmore, however, the foremost evidence scholar of our time, feels that re-
stricting the rule to statements of deceased persons is the sounder approach:

It would seem, however, that *insanity,* or *absence* from the jurisdiction, would here not suffice (as it does for some of the foregoing
exceptions); because the necessity in general is predicated on titles
and boundaries of long standing, for which the lapse of time has op-
erated to destroy other evidence; and hence if the matter is one of the
present generation, or if the evidence in question comes from the present generation (as it would if the declarant were merely absent), this
necessity can hardly be presumed to exist.[57]

---

53. *Wood v. Willard,* 37 Vt. 377, 387 (1864).
54. *Caufman v. Prebyterian Congregation of Cedar Spring,* 6 Binn. 59, 62 (Pa. 1813).
55. *Sasser v. Herring,* 14 N.C. (3 Dev.L) 340, 342 (1832); *see also Barrett v. Kelly,* 131 Ala. 378,
30 So. 824 (1901); *O'Connell v. Cox,* 179 Mass. 250, 60 N.E. 580 (1901).
56. Cal. Evid. Code § 1323.
57. 5 Wigmore on Evidence § 1565, at 515 (rev. ed., 1974).

The general requirement that hearsay evidence, if it is to be received at all, bear a "circumstantial probability of trustworthiness" requires that the declarant appear to have had no interest in misrepresenting the truth. Accordingly, in many jurisdictions, statements of an owner about his own boundaries are inadmissible.

> [I]t must be presumed to have been their interest to extend the boundaries of the lot, and their declarations in favor of their interest were clearly not evidence.[58]

California, on the other hand, in common with several other states, does not automatically disqualify the statements of an owner. The California statute quoted above goes on to provide the following: ". . . but evidence of a statement is not admissible under this section if the statement was made under circumstances such as to indicate its lack of trustworthiness."[59]

In a curious twist on this rule, Massachusetts and several other states *required* that such declarations be made by an owner in possession, and while he is on the land and in the act of pointing out the boundaries in question.[60]

The third requirement, developed from the general principle of evidence requiring personal knowledge of a witness, is that the declarant must be shown to have knowledge of the location of the boundary in question. The California rule quoted above, for example, requires that the declarant "had sufficient knowledge of the subject. . . ."[61] An early Virginia case held the following:

> [Declarations respecting boundaries are admissible] provided such person had peculiar means of knowing the fact; as, for instance, the surveyor or chain carrier who were [sic] engaged upon the original survey, or the owner of the tract, or of an adjoining tract calling for the same boundaries; and so of tenants, processioners, and others whose duty or interest would lead them to diligent enquiry and accurate information of the fact. . . .[62]

The Federal Rules of Evidence do not provide an explicit exception to the hearsay rule for individual statements concerning boundaries. However, such statements could probably be admitted under the broad, general exception contained in Federal Rule 803(24), discussed below.

---

58. *Shepherd v. Thompson*, 4 N.H. 213, 215 (1827).

59. Cal. Evid. Code § 1323.

60. *Goyette v. Keenan*, 196 Mass. 416, 82 N.E. 427(1907). *See also Emmet v. Perry*, 100 Me. 139, 141, 60 A. 872, 873 (1905); *Curtis v. Aaronson*, 49 N.J.L. 68, 7 A. 886 (1887); *Collins v. Clough*, 222 Pa. 472, 71 A. 1077 (1909); *Hunnicutt v. Peyton*, 102 U.S. 733 (1880); *Clement v. Packard*, 125 U.S. 309, (1887); *Ayres v. Watson*, 137 U.S. 584 (1891); *Robinson v. Dewhurst*, 68 F. 336 (4th Cir. 1895).

61. Cal. Evid. Code § 1323; *see also Morton v. Folger*, 15 Cal. 275 (1860); *Morcom v. Baiersky*, 16 Cal.App. 480, 117 P. 560 (1911).

62. *Harriman v. Brown*, 35 Va. (8 Leigh) 697, 713 (1837).

RECITALS IN DEEDS AND OTHER WRITINGS AFFECTING PROPERTY

Recitals contained in deeds and other documents affecting interests in real property are admitted as an exception to the hearsay rule in some jurisdictions under limited circumstances. In California, for example, a recital in a deed, will, lease, and so on, is admissible if:

(a)   The matter stated was relevant to the purpose of the writing;

(b)   The matter stated would be relevant to an issue as to an interest in the property; and

(c)   The dealings with the property since the statement was made have not been inconsistent with the truth of the statement.[63]

Such recitals in California may take on a good deal of evidentiary significance. California Evidence Code section 622 provides: "The facts recited in a written instrument are *conclusively* presumed to be true as between the parties thereto, or their successors in interest; but this rule does not apply to the recital of a consideration." [Emphasis added.] (For the effect of a conclusive presumption, see the chapter on presumptions, Chapter 6,) Similarly, the Federal Rules provide an exception for:

[a] statement contained in a document purporting to establish or affect an interest in property if the matter stated was relevant to the purpose of the document, unless dealings with the property since the document was made have been inconsistent with the truth of the statement or the purport of the document."[64]

Thus, hearsay evidence of this type may not only be admitted as an exception to the general rule, but may also conclusively establish the recited facts.

ANCIENT DOCUMENTS

An exception to the hearsay rule found in virtually all jurisdictions applies to what are, for convenience, called "ancient documents." Typically, if a writing is more than 30 years old and contains nothing that casts suspicion on its reliability, and if the writer is shown to have been a qualified witness (e.g., having firsthand knowledge), the writing may be received for the truth of the matter asserted within it.

Originally, courts invoked the factors of age and lack of suspicion solely for the purpose of authenticating documents. The difficulty of obtaining witnesses to testify to the authenticity of an "ancient document" was considered a sufficiently compelling reason to allow the writing to "self-authenticate." But the document was not received for the truth of the matter asserted in it unless it qualified under an exception to the hearsay rule. Today, a large number of American courts nevertheless have extended the principle to the realm of the hearsay rule, holding that such ancient documents have an exception of their own. McCormick has questioned whether such an extension of the rule is supported in principle and has concluded that

---

63.   Cal. Evid. Code § 1330.
64.   Fed. R. Evid. 803 (15).

it is.[65] He notes that because the exception is limited to writings and does not apply to oral statements made many years earlier, there is not the danger that the statement cannot be retold accurately. Furthermore, the age requirement helps to ensure that the statement was made at a time before the present controversy arose and consequently that the declarant had no motive to misrepresent the truth. What is more, there is a further assurance of reliability in the requirement that the witness be shown to have had the usual qualifications of a witness; he must be shown to have had firsthand knowledge, for example.

Not surprisingly, the majority of cases treating this exception to the hearsay rule have done so when considering the admissibilty of plats, field notes, and other writings in cases of land title and boundaries. In *Plattsmouth Bridge Co. v. Globe Oil and Refining Co.*, 232 Iowa 1118, 7 N.W.2d 409 (1943), the court upheld the admission into evidence of an ancient plat made by a county surveyor, writing that "[i]t was properly identified, is more than thirty years old, and was competent as an ancient document." In *Lawrence v. Tennant*, 64 N.H. 532, 15 A. 543 (1888), the court admitted a plot of the town of Epsom, "an ancient and much-worn document" purporting to have been done in the year 1800 by L. D. Morrill. "The antiquity and genuineness of the plan thus appearing, it was, of course, admissible if made by public authority; and, if not so made, it was none the less admissible. . . ."[66]

The federal and California rules for ancient documents are structured somewhat differently from these general principles. Federal Rule of Evidence 803(16) provides that "[s]tatements in a document in existence 20 years or more the authenticity of which is established" are admissible as an exception to the hearsay rule. The question of the authenticity of the ancient document is separately treated in Rule 901(b)(8). That rule provides that authentication is established by:

> Evidence that a document or data compilation, in any form,
> (A) is in such condition as to create no suspicion concerning its authenticity, (B) was in a place where it, if authentic, would likely be, and (C) has been in existence 20 years or more at the time it is offered.

The California rule contains a requirement not imposed by other jurisdictions:

> Evidence of a statement is not made inadmissible by the hearsay rule if the statement is contained in writing more than 30 years old and the statement has been since generally acted upon as true by persons having an interest in the matter.[67]

The California Law Revision Commission, which drafted the California Evidence Code (the Code was enacted in 1965), noted that the ancient-documents concept had been extended from the area of authentication to the hearsay rule. The Commission's view was that the age of a document alone is not a sufficient guarantee of the trustworthiness of the statement

---

65. McCormick on Evidence § 323, at 747 (2d ed., 1972).

66. *Lawrence v. Tennant*, 15 A. 543, 545; for a collection of other cases, *see* Annot., 46 A.L.R.2d 1318 (1956).

67. Cal. Evid. Code § 1331.

contained in it. For that reason it added the requirement that the statement must have been generally acted upon as true by people having an interest in the subject.

Cases illustrative of the ancient documents exception include *City of New York v. Wilson & Co.,* 278 N.Y. 86, 15 N.E.2d 408 (1938), an ejectment action brought by the City of New York to recover possession of land lying waterward of the original high-water line of the East River. The New York Court of Appeals held that under that state's statute, maps that had been filed more than 20 years were admissible on the city's contention that the original high-water line was landward of the presently existing line. In another ejectment action, the court in *State v. Taylor,* 135 Ark. 232, 205 S.W. 104, 105 (1918), held that plat books, together with notations in them made at the time the plats were made, were admissible as ancient documents. The court pointed out that "[o]bservation of the entries on this record show it to be an ancient one, free from suspicion, and made at a time when the officer had information before him with reference to which the notations were made." In *Hart v. Gage,* 6 Vt. 170, 172 (1834), the Vermont Supreme Court held certain maps and field books admissible as ancient documents, commenting that "these books, and the maps made from them, as they ripen by time, and monuments perish, may, like Doomsday Book, be the best, if not the only evidence of many ancient surveys."[68]

An example of the requirement that the writing be free from suspicion is *Muehrcke v. Behrens,* 43 Wis.2d 1, 169 N.W.2d 86 (1969). In that case a 37-year-old town record book was sought to be introduced to show that the town board had regarded certain land as a public highway. The legal description of the land in question referred to a length of "80" rods, over which the number "160" had been written. The court held that when an instrument has been altered, the party who seeks its introduction must explain the alteration. Ultimately, the court concluded that the book was admissible because other evidence corroborated the fact that the distance in question was 160 rods.

The requirement that the writer be shown to have had personal knowledge of the assertion contained in the document is illustrated by *Budlong v. Budlong,* 48 R.I. 144, 136 A. 308 (1927). In that case, a book found in the office of a poor farm contained an entry suggesting that a child was born to a certain woman and was offered to prove the identity of the mother. The court held the book inadmissible because there was no proof that the writer had personal knowledge of the facts stated; in fact, the identity of the writer was not shown at all.

As in the case of community reputation, some courts receive ancient documents on questions of boundaries, but not of title. Thus, in *Jackson, Ex dem. Beekman v. Witter,* 2 Johns 180 (New York 1807), the court held inadmissible as evidence of title a map made purportedly by commissioners appointed to partition certain lands: "Though it is a very ancient transaction, and might be good evidence as to the boundary or location of the premises, it never can be considered as conveying any title."[69]

---

68. Other illustrative cases include *Spencer v. Levy,* 173 S.W. 550 (Tex.Civ.App. 1914); *Lowell v. Boston,* 322 Mass. 709, 79 N.E.2d 713, app. dism'd. 335 U.S. 849 (1948); *McCausland v. Flemming,* 63 Pa. 36 (1869); *Gibson v. Poor,*125 21 N.H. 440, 53 Am. Dec. 216 (1850); and, *Taylor v. McConigle,* 120 Cal. 123, 52 P. 159 (1898).
69. The same principle was applied in *Finberg v. Gilbert,* 104 Tex. 539, 141 S.W. 82 (1911), and *The Schools v. Risley,* 77 U.S. 91 (10 Wall.) (1869).

FEDERAL RULE OF EVIDENCE 807

After listing, in Rules 803 and 804, some 29 specific exceptions to the rule against hearsay, Rule 807 in the end provides a general exception for hearsay having the basic attributes of trustworthiness and need:

> [A] statement not specifically covered by any of the foregoing exceptions but having equivalent circumstantial guarantees of trustworthiness, if the court determines that (A) the statement is offered as evidence of a material fact; (B) the statement is more probative on the point for which it is offered than any other evidence which the proponent can procure through reasonable efforts; and (C) the general purposes of these rules and the interests of justice will best be served by admission of the statement into evidence. However, a statement may not be admitted under this exception unless the proponent of it makes known to the adverse party sufficiently in advance of the trial or hearing to provide the adverse party with a fair opportunity to prepare to meet it, the proponent's intention to offer the statement and the particulars of it, including the name and address of the declarant.

This provision plainly vests the trial judge with a substantial discretion to receive hearsay statements into evidence. But does it portend the death of the rule against hearsay in federal courts? Probably not. Few decisions have treated Rule 803(24), but one federal court has declared it should be used sparingly: "It was the intent of Congress that this exception be used rarely and only in exceptional circumstances."[70] Perhaps more significantly, the trial bar will not idly watch its hard-gotten knowledge of the arcane hearsay rule rendered valueless by a rule having such a common-sense ease of application. At any rate, this new residual exception to the hearsay rule has, so far, been little used.

As a parting remark, it should be mentioned that the common-law's fastidiousness about hearsay evidence is undergoing a crucible of self-examination in many of the nations—other than the United States—that share the English common-law heritage.[71]

---

70. *Lowery v. Maryland*, 401 F.Supp. 604, 608 (D. Md. 1975); *but see United States v. Iaconetti*, 406 F.Supp. 554 (E.D. N.Y. 1976); *Muncie Aviation Corp. v. Party Doll Fleet, Inc.*, 519 F.2d 1178 (5th Cir. 1975).

71. While not in the United States, in many other nations of the common-law heritage the hearsay rule is coming in for a rough go. *See, for example*, in Canada: Ontario Law Reform Commission "Report on the Law of Evidence" (1976) 14–15; in Australia: Law Reform Commission of New South Wales "Report on the Law of Hearsay" (LRC 29, 1978) para. 6.12.8; in Ireland: Law Reform Commission (WP 9, 1980) 28 adopted by reference and without amendment in the relevant respect in the final "Report on the Rule Against Hearsay in Civil Cases" (LRC 25, 1988); in Scotland: Scottish Law Commission, "Evidence: Report on Corroboration, Hearsay and Related Matters in Civil Proceedings" (Scot. Law Com. No. 100, 1986) para. 3.32; and in New Zealand: New Zealand Law Commission "Evidence Law: Hearsay" (Preliminary Paper No. 15, 1991) paras. 48–49. And see Ulrich, "Reform of the Law of Hearsay" (1974) Anglo-Am LR 184, 209; R. F. G. Ormrod, "Evidence and Proof: Scientific and Legal" (1972) 12 Med. Sci. & L. 9 16; J. D. Heydon, *Evidence, Cases and Materials,* 2nd ed. (London: Butterworths, 1984) 361–362; and English Law Commission, *The Hearsay Rule in Civil Proceedings* (Consultation Paper No. 117, 1991) para. 3.50.

Chapter 4

The Rule Requiring Personal Knowledge

"Do we ever hear the most recent fact related exactly in the same way, by the several people who were at the same time eyewitnesses to it? No. One mistakes, another misrepresents; and others warp it a little to their own turn of mind, or private views."

—Lord Chesterfield, letter to his son, April 26, 1748, in *I Letters of Chesterfield* (London: Swan Sonnersehen & Co., Ltd., 1905), p. 105

Thus far it should be apparent that no matter how contrived or obscure the rules of evidence may seem, they are unmistakably the product of the steadfast purpose of the common law and American courts to receive only the most reliable evidence. This purpose also accounts for the fundamental rule requiring that witnesses testify only to matters of which they have personal knowledge.[1] A customary statement of the rule is that a witness who testifies to a fact that could have been perceived by the senses must have had an opportunity to observe and must also have actually observed the fact.[2] Federal Rule of Evidence 602, for example, provides the following:

> A witness may not testify to a matter unless evidence is introduced sufficient to support a finding that he has personal knowledge of the matter. Evidence to prove personal knowledge may, but need not, consist of the testimony of the witness himself. This rule is sub-

---

1. The obverse of the rule requiring personal knowledge is one of the older common law rules—that is, that witnesses generally may not testify as to their opinions. *See, generally,* 2 Wigmore on Evidence §§ 650–670; 97 C.J.S. *Witnesses* §52 (1957); 58 Am. Jur. *Witnesses* §§ 113, 114 (1971).
2. For representative cases, *see Barnett v. Aetna Life Ins. Co.,* 139 F.2d 483 (3rd Cir. 1943); *State v. Dixon,* 420 S.W.2d 267 (Mo. 1967); *State v. Johnson,* 92 Idaho 533, 447 P.2d 10 (1968).

ject to the provisions of Rule 703, relating to opinion testimony by expert witnesses.

California Evidence Code section 702 provides the following:

(a)   Subject to Section 801 [relating to opinion testimony by expert witnesses], the testimony of a witness concerning a particular matter is inadmissible unless he has personal knowledge of the matter. Against the objection of a party, such personal knowledge must be shown before the witness may testify concerning the matter.

(b)   A witness' personal knowledge of a matter may be shown by any otherwise admissible evidence, including his own testimony.

The California statement of the rule is different from the federal rule in one material respect. Ordinarily a judge may in his discretion allow the admission of evidence without a showing of a required foundational fact—in this case the personal knowledge of the witness—if the propounder of the evidence represents that he can provide it later in the course of the trial. If the propounder is subsequently unable to produce the foundational fact, the evidence is subject to be stricken from the record. The drafters of the California Evidence Code apparently felt that the rule requiring firsthand knowledge is of such importance that the judge should have no discretion, but must require a showing of the foundational fact before allowing the evidence to be received. As the drafters noted in their comment to section 702: "If a timely objection is made that a witness lacks personal knowledge, the court may not receive his testimony subject to the condition that evidence of personal knowledge be supplied later in the trial. Section 702 thus limits the ordinary power of the court with respect to the order of proof."[3]

The essential concept of the rule requiring personal knowledge is that, with the exception of opinion testimony given by experts, a witness may testify only to facts, and only when he has personal knowledge of those facts. Thus we find the Supreme Court of Alabama in 1848 writing the following:

The general rule requires, that witnesses should depose only to facts, and such facts too as come within their knowledge. The expression of opinions, the belief of the witness, or deductions from the facts, however honestly made, are not proper evidence as coming from the witness; and when such deductions are made by the witness, the prerogative of the jury is invaded.[4]

As mentioned above, the requirement that witnesses have personal knowledge of what they testify to is an ancient one. In 1349, it was held that witnesses, as opposed to jurors, were not

---

3.   California Law Revision Commission comment to Evidence Code Section 702, West's Annotated California Evidence Code, p. 116.

4.   *Donnell v. Jones,* 13 Ala. 490, 510 (1848). In the *Donnell* case, the court held that the opinion of a witness familiar with business matters, whether a levy of attachment had destroyed the credit of a business, was held properly excluded.

challengable "because the verdict will not be received from them, but from the jury; and the witnesses are to be sworn to 'say the truth,' without adding 'to the best of their knowledge,' for they should testify nothing but what they know for certain, that is to say what they see and hear."[5]

The foundational requirement—to establish that the witness has personal knowledge of what he is about to testify to—is a burden of the party offering the testimony. Unlike California, most states allow the trial judge discretion to admit the evidence, deferring the proof of personal knowledge to a later point in the trial.[6] If the adversary does not promptly object when evidence is offered without a showing of personal knowledge, he has waived his right to object to the admissibility of the evidence. But if he objects and later shows that the witness did not have personal knowledge, he will be entitled to have the evidence in question stricken.[7]

In establishing the foundation of personal knowledge, the examining attorney may inquire as to the circumstances under which the witness had occasion to learn of the matters he is about to testify to. This accounts for such questions as, "How is it that you recall that the time was 10:05?," or "How can you be sure that the date in question was October 26, 1967?" Such questions, when plausibly answered by the witness, can do much to enhance his credibility. For example: "I was late for a 9:30 meeting, and I had been checking my watch every 5 minutes," or "That date also happens to be my wedding anniversary."

Occasionally, a shrewd attorney may purposely neglect to ask a question that would best serve to demonstrate that his witness has personal knowledge of the fact, leaving it to his opponent to blunder into it on cross-examination and thus serve to enhance the credibility of the witness far more than could the attorney who called him. That is what occurred in the oft-told tale of the witness in a mayhem case who had testified that the defendant bit off the ear of the victim. On cross-examination, the opposing counsel succeeded in establishing that the witness had not actually seen defendant do the alleged biting, nor appeared to have had any other means of perceiving the event. Rather than be content with his answers, the attorney at last foolishly asked, "Well then, just how is it that you know the defendant bit off the ear of the victim?" The witness replied, "I saw him spit it out."[8]

The rule requiring personal knowledge is frequently confused with the rule against hearsay, but the two have different origins, as has been noted, and frequently different applications. If the testimony of a witness lacks a showing that the witness had firsthand knowledge of the event—without showing that his testimony was based on the statements of others— the evidence is objectionable on the ground that the witness lacked firsthand knowledge. If, on the other hand, the same testimony shows that its basis is the statements of others, both objections are availing.

As in the case of the rule against hearsay, the law has made exceptions to the rule requiring firsthand knowledge. Although not nearly so formally categorized as in the case of the hearsay rule, where logic, the presence of other indicia of reliability, and sheer need for the testimony are present, exceptions to the rule requiring personal knowledge have been formulated.

---

5.   Anon. Lib. Assn. 110, 11 (1349), quoted in Phipson, Evidence 398 (19th ed., 1952).
6.   See, e.g., Sofas v. McKee, 100 Conn. 541, 124 A. 380 (1924).
7.   See, e.g., State v. Dixon, 420 S.W.2d 267 (Mo. 1967).
8.   The wise and ancient dictum to the young trial lawyer holds that he ought never, on cross-examining a witness, to ask a question that begins "Why . . .?," particularly when he does not already know the answer.

It may also be said that some of these exceptions have been created to prevent ludicrous results. For example, witnesses are invariably allowed to testify to their age and date of birth.[9] Similarly, witnesses have been allowed to testify as to their kinship to others.[10] The same experience holds true when the required evidence as a practical matter can be found only in a body of records, whether public or private. It is simply impractical, if not impossible, in such cases to require the summoning to the witness stand of each person having personal knowledge of the events recorded in such records. Thus in one case a witness was allowed to testify to the data shown on two graphs compiled from records of his water company.[11] In another, the clerk of an Adjutant General's office having more than 20 years' service was allowed to testify to the practices of his office prior to his employment.[12] In another, testimony of the fact and date of the filing of surveyor's notes was allowed, although the witness had no personal knowledge of these facts and his knowledge was based primarily on the mere presence of the records in the office.[13] In another decision, the requirement of personal knowledge of the contents of land-title abstracts and other land records, made before the Chicago fire of 1871, was dispensed with.[14]

Similarly, testimony respecting data collected from the use of scientific instruments, formulas, and tables invariably entails a dependence on the statements of other persons, and even of unknown inventors, computer programmers, and so on. Yet it is clearly impracticable to adhere rigidly to the firsthand knowledge requirement when the sole source of the necessary firsthand knowledge "connection" cannot feasibly be produced.

The burden of the rule requiring firsthand knowledge is made more acute when a party seeks to prove the *nonexistence* of an event. How can a witness be shown to have personal knowledge that someone never existed, or that a document was never executed? This problem is pondered in greater detail in the chapter dealing with the rule against hearsay (Chapter 3), but the remarks of one thoughtful judge may give the reader a glimpse of the evidentiary problem (though not, perhaps, of its resolution). A 1945 Pennsylvania decision considered the testimony of a witness who testified that "no whistle was blown and no bell was rung." While the content of the case is of no moment for present purposes, the decision did afford Chief Justice Maxey the opportunity to make these observations on the matter of human perceptions:

> It is impossible to lay down any rule by which it can infallibly be determined that a witness' statement that "no whistle was blown and no bell was rung" is of so positive a character as to ·support a charge of negligence, or is of such negative character as to "amount only to a scintilla." . . .

---

9.   *See, e.g., Antelope v. United States,* 185 F.2d 174 (10th Cir. 1950); *Hancock v. Supreme Council Catholic Benevolent Legion,* 69 N.J.L. 308, 55 A. 246 (1903) [witness allowed to testify to the age of an elder brother].

10.   *Brown v. Mitchell,* 88 Tex. 350, 31 S.W. 621, 623 (1895) [the witness was allowed to testify that he was the child of the deceased, notwithstanding that he lacked firsthand knowledge of the sole event which could have provided that knowledge].

11.   *Bridgeport Hydraulic Co. v. Town of Stratford,* 139 Conn. 388, 94 A.2d 1, 5 (1953).

12.   *Worcester v. Northborough,* 140 Mass. 397, 402, 5 N.E. 270 (1886).

13.   *Hill v. Kerr,* 78 Tex. 213, 14 S.W. 566 (1890).

14.   *Chicago & A. R. Co. v. Keegan,* 152 Ill. 413, 39 N.E. 33 (1894).

Two persons of equally keen hearing may be sitting in a room, and one of them will hear a noise outside while the other will not. A person habituated to intense mental concentrations is not as likely to hear noises as is one whose mind is more objective in character. It is perhaps correct to say that extroverts are more perceptive of sights and sounds than are introverts. The mind of the one is directed outward; the mind of the other is directed inward. Very frequently an introvert does not *see* an object he *looks at.* Two persons equal in vision may be sitting side by side and looking in the same direction and one of them will see objects to which the other is perfectly blind. The same phenomena often obtains [*sic*] with two persons of equal auditory powers; one will hear because his entire interest is in externals; the other will not hear because of his propensity for reflection.[15]

It may seem surprising that a rule at least as old as the hearsay rule, and one seeming to require for efficient judicial administration at least as many exceptions, should find so little treatment in the cases, and for its exceptions so few compartments. For analytical purposes, an in-depth study of the exceptions that have been made to the personal-knowledge rule might show that each exception could alternatively be viewed as an admission of opinion testimony. After considering the material on the opinion rule and the meaning of the expression "expert witness," the reader may wish to ponder whether the witnesses in the preceding cases making "exceptions" to the personal knowledge rule might as well be called "expert witnesses," and their testimonies "opinions." On the other hand, because so much of the testimony that is subject to the firsthand knowledge objection is also subject to a hearsay objection, it may be that the courts have contented themselves with the exceptions they have fashioned for the latter rule.

---

15. *Kindt v. Reading Co.,* 352 Pa. 419, 424, 43 A.2d 145, 147–148 (1945). *See also* Wigmore, *Science of Judicial Proof, as Given by Logic, Psychology, and General Experience, and Illustrated in Judicial Trials,* §§ 191–208 (3d ed., 1937).

Chapter 5

Of Proof and Other Burdens

"Heed must be given to the burden of proof, at least when other tests are lacking."

—Justice Benjamin Cardozo, in *Shapleigh v. Mier,* 299 U.S. 468, 475(1937)

The expression "burden of proof" conveys an idea that may be instinctively understood—that in an adversary proceeding, one side must bear the onus of proving that its position is correct. Put another way, if the evidence on each side of a case is equally persuasive, in "equipoise," one side nevertheless must prevail. In championship prize fights, the challenger must take the crown from the champion. If they draw, the title holder keeps his title.

The common law has refined the concept of burden of proof to a remarkable sophistication, on the order of calculating the number of angels pirouetting on the head of a pin. As refined, the expression "burden of proof" actually denotes two distinct burdens, for convenience called the "burden of producing evidence" and the "burden of persuasion." The burden of producing evidence is sometimes referred to as the "burden of production," or the "duty of going forward."[1] The burden of persuasion is occasionally called the "risk of non-persuasion"[2] or, unhappily, for the confusion it creates, the "burden of proof."[3]

A.   THE BURDEN OF PRODUCTION

The burden of production is best understood in the context of a jury trial. If one party has the burden of production on an issue (for example, whether a particular survey was fraudulent), he must produce sufficient evidence from which it might reasonably be con-

---

1.   McCormick on Evidence § 336, at 783, fn. 3 (2d ed., 1972). Interestingly, assigning the task of producing evidence to the parties, rather than to the judge, is a peculiar characteristic of the common law of England not found in the civil law systems of continental Europe. 9 Wigmore on Evidence § 2483, at 266–267 (3d ed., 1940).
2.   9 Wigmore on Evidence § 2485, at 271 (3d ed., 1940).
3.   Cal. Evid. Code § 500.

cluded that the survey was in fact fraudulent. If he does not, the judge may direct a finding against him and refuse to allow the jury to decide the issue. If the issue is essential to the party's entire case, of course, he then loses. When he has introduced sufficient evidence— say, the expert testimony of a surveyor who has analyzed the survey and concluded it was fraudulent—the burden is discharged. (The party may still be required to carry the burden of persuasion, however, which is discussed below.)

The other party during the presentation of his case may then produce evidence that the survey was not fraudulent or he may produce no evidence on the point, being content merely to argue against the persuasiveness of his opponent's evidence. On occasion, however, the second party *must* produce evidence of his own or risk a directed finding against him. This circumstance is termed a "shifting" of the burden of production and occurs in several situations.

One such situation occurs when the evidence produced by the party initially having the burden is so persuasive in proving a fact that, in the opinion of the trial judge, reasonable men could not without other evidence conclude otherwise than that the fact is established. Suppose the issue is whether Jones was in a tavern on a particular night, and the party having the burden of production produces three witnesses who testify to Jones's presence there. So long as the witnesses do not appear to be lying, the judge may cause the burden of production to shift to the other party; in other words, if the second party produces no evidence that Jones was *not* in the tavern, the judge will direct the jury to find that he was. To take another example, suppose the plaintiff has the burden of production on his assertion that the elevation of mean high water at Crescent City, California is 4.20 feet above N.G.V.D.[4] He produces an official of the National Ocean Survey, who testifies about his agency's authority and also about the fact that the Survey has installed a tide gauge at Crescent City and that from the gauge's readings the Survey has computed the value of mean high water to be 4.20 feet above N.G.V.D. In such a circumstance, if the opponent produces no evidence showing the value to be different, the judge will likely not allow the jury to find a different value for the datum.

In *People v. Ward Redwood Co.*, 225 Cal.App.2d 385 (1964), the State claimed title to land known as Taylor Island. The State's theory was that as of the date of California's statehood (1850), the land lay below the ordinary high-water mark of the Klamath River and thereafter emerged from the riverbed as an island. At the conclusion of the State's case, the defendant presented no evidence, apparently in the belief that the State had not met its burden of *persuasion*. Judgment was entered, however, for the State. On appeal, the court examined the evidence produced by the State and, finding it sufficiently persuasive, held that the "burden" had shifted to the defendant.[5] Because the defendant had not met his burden when it had shifted to him, the judgment for the State was affirmed.

What factors determine who bears the burden of production on a given issue? Generally, the party who must plead a fact has the burden of producing evidence to prove the fact. In an ejectment action, the plaintiff must plead, and therefore bear the burden of producing evidence, that he is the owner of the subject property. If during his case the plaintiff neglects to present evidence that he owns the property (concentrating instead, for example, on a

---

4.   That is, the National Geodetic Vertical Datum. *See* Variability of Tidal Datums and Accuracy in Determining Datums from Short Series of Observations, NOAA Techinical Report NOS 64 at 4–5 (1974).
5.   Only the burden of production can "shift"; hence it was this burden that the court referred to.

weakness on defendant's claim of title), he will fail to meet his burden of production. Thus, the universal rule has arisen that in title actions the plaintiff must prevail on the strength of his own title and may not solely rely on the weakness in defendant's title.[6]

If the defendant pleads in defense that the plaintiff is estopped to prove his title, [7] or that the plaintiff's claim is barred by a statute of limitations,[8] he bears the burden of production on that issue. If the defendant in such a case neglects to produce any evidence on such issues of estoppel or limitations, the judge will direct a finding against him on those issues, and the jury will not be allowed to consider the defenses. In a judge-tried case, the judge will find for the plaintiff on such an issue.

If a party superfluously pleads a matter he is not required to plead, the prevailing view is that he will not thereby incur the burden of producing evidence on the matter.[9]

It must be emphasized, however, that the burden of *pleading* does not in all cases determine the burden of producing evidence. Other factors for allocating the burden of production may be found in the cases, some sound, some perhaps not so. A frequently encountered principle is that when the fact of the matter is more apt to be known by one party, it is he who should bear the burden of production. In an action on a debt, for example, it is thought that the defendant is more likely than the plaintiff to have evidence that he has paid the debt; once the plaintiff establishes the debt, accordingly, it is incumbent on the defendant to show he has paid it.[10] Another consideration in fixing the burden of production is the likelihood of the fact. Where services are performed for a family member, they are more likely to be a gift than when they are performed for a regular customer.[11] The burden of production, accordingly, may be allocated to the party urging that the services were not intended as a gift.

It is sometimes stated that a party must prove a fact he pleads, unless his averment (the pleaded fact) is negative in form:

> The first rule laid down in the books on evidence is to the effect
> that the issue must be proved by the party who states an affirmative,
> not by the party who states the negative.[12]

The better and prevailing view, however, is that the form of averment, whether affirmative or negative, should not be determinative in assigning the burden of production. Certainly all propositions can be stated in a negative form. Thus in *Johnson v. Johnson,* 229 N.C. 541, 50 S.E.2d 569 (1948), the plaintiff alleged in his reply to the defendant's answer that the deed in question had not been executed by its purported grantor; he was nonetheless given the burden of production on the question.

---

6.   *United States v. Oregon,* 295 U.S. 1 (1935); *Ernie v. Trinity Lutheran Church,* 51 Cal.2d 702, 706 (1959); *Helvey v. Sax,* 38 Cal.2d 21, 23 (1951); *see* cases cited in 74 C.J.S. *Quieting Title* § 17, at 41, fn. 58 (1951).

7.   *See, e.g., City of Long Beach v. Mansell,* 3 Cal.3d 462 (1970).

8.   *See, e.g., California v. Arizona,* 440 U.S. 59, 63, fn. 4 (1979).

9.   McCormick on Evidence § 337, at 785, fn. 12 (2d ed. 1972).

10.   *See, e.g.,* Fed. R. Civ. P. 8(c).

11.   James, Civil Procedure 257 (1965).

12.   *Walker v. Carpenter,* 144 N.C. 674, 676, 57 S.E. 461 (1907); *See also Levine v. Pascal,* 94 Ill. App.2d 43, 236 N.E.2d 425 (1968).

B. THE BURDEN OF PERSUASION

The burden of persuasion is the burden on a party to "prove" (that is, convince the trier of fact) the existence or nonexistence of a fact essential to his claim for relief or to his defense. The plaintiff in a quiet-title action, for example, claims to own the disputed property. It is incumbent upon him to convince the trier of fact that he does; he may not rely simply on the defendant's inability to show title in himself. Likewise, a defendant who asserts in defense that a deed to the plaintiff was not executed by the purported grantor, but was forged, has the burden of convincing trier of fact of his point.

The burden of persuasion is quite different from the burden of production. Unlike the burden of production, the burden of persuasion is not assigned until all the evidence has been received.[13] Moreover, the burden of persuasion does not "shift" during the course of the trial (except in the unusual case of a *presumption* causing it to shift; *q.v.,* below). There is one similarity, however; The burden of persuasion is allocated by the same principles that allocate the burden of production in the first instance (that is, before any "shift" in that burden).[14]

C. THE QUANTUM OF PROOF

The word "prove" two paragraphs above was placed within quotation marks advisedly, because it is a protean word—it means different things in different contexts. In general, it means that the party having the burden of persuasion on an issue must induce a *particular degree* of conviction in the mind of the trier of fact (the jury or, when there is no jury, the judge). If the required degree of conviction is not achieved, the trier of the fact must assume that the fact does not exist.[15] The degree, or quantum, of proof varies. In criminal cases, the prosecution must prove each element of the crime charged "beyond a reasonable doubt."[16] When a criminal defendant pleads an affirmative defense, such as duress or insanity, the degree of proof he must achieve to meet his burden of persuasion varies from state to state. The United States Supreme Court has held that it is constitutional for a state to require a defendant to prove such an affirmative defense beyond a reasonable doubt.[17] The expression "beyond a reasonable doubt" ought to be readily understandable. One judge has written that it requires a remarkable skill in language to make it plainer by explanation.[18]

In civil actions, on the other hand, the degree of proof required of the party bearing the burden of persuasion is ordinarily "the preponderance of the evidence." To prove a matter by a preponderance of the evidence is not simply to produce more witnesses or introduce more

---

13. McCormick on Evidence § 337, at 788 (2d ed., 1972).
14. McCormick on Evidence § 337, at 788 (2d ed., 1972).
15. Morgan, Basic Problems of Evidence 19 (1954). *cf.* "The Murders in the Rue Morgue," *The Complete Tales and Poems of Edgar Allan Poe* (New York: The Modern Library 1965), p. 159. Mr. Dupin says, "You will say, no doubt, using the language of the law, that to make out my case, I should rather undervalue than insist upon a full estimation of the activity required in this matter. This may be the practice in law, but it is not the usage of reason. My ultimate object is only the truth."
16. *In re Winship,* 397 U.S. 358, 364 (1970).
17. *Leland v. Oregon,* 343 U.S. 790 (1952).
18. Newman, J., in *Hoffman v. State,* 97 Wis. 571, 73 N.W. 51, 52 (1897).

documents than the opponent.[19] Simply stated, this burden of persuasion is met when a party produces evidence (together with inferences that the jury can reasonably make from the evidence) that is more convincing than his opponent's evidence. This concept has been treated in rarified detail elsewhere.[20] And so this discussion should suffice for present purposes.

"Clear and convincing evidence" is a measure or a quantum of proof that lies somewhere between proof by a preponderance of the evidence and proof beyond a reasonable doubt. The phrasing of this measure of proof varies somewhat from state to state, and it has been cogently suggested that the expression means simply this: The trier of fact must be convinced that the fact to be proven is highly probable.[21] Clear and convincing proof is required of the party bearing the burden of persuasion in a variety of civil cases. Examples are suits alleging an oral contract to make a will,[22] suits to establish the terms of a lost will,[23] actions to reform or modify written transactions such as deeds or contracts,[24] and so forth. It is difficult to derive a principle that states accurately when clear and convincing proof is required in civil actions instead of proof by a preponderance of the evidence. In general, however, the greater measure of proof is required when a party's contention strongly belies common-sense experience or when the contention would avoid a strong matter of policy. Thus, in California, for example, the owner of the legal title to real property is presumed to be the owner of the full beneficial title, and this presumption is rebuttable only by "clear and convincing proof."[25] A mortgage or deed of trust would constitute such clear and convincing evidence, of course, and the rule seeks to ensure that very little else could upset the presumption. By the same token, the law strongly favors written wills to provide certainty and reduce disputes in the disposition of property after death. Accordingly, one who asserts that he had an oral agreement whereby the deceased was to leave him certain property bears this heavier burden of persuasion.

---

19. *See Livingston v. Schreckengost,* 255 Iowa 1102, 125 N.W.2d 126, 131 (1963).

20. *See Burch v. Reading Co.,* 240 F.2d 574 (3d Cir. 1957), *cert. den.* 353 U.S. 965 (1957); *Sargent v. Massachusetts Accident Co.,* 29 N.E.2d 825, 827 (1940); *McDonald v. Union Pac. R. Co.,* 109 Utah 493, 167 P. 2d 685, 689 (1946); 9 Wigmore on Evidence § 2498, at 326 (3d ed., 1940).

21. McBaine, *Burden of Proof: Degrees of Belief,* 32 Calif.L.Rev. 242 (1944).

22. *See, e.g., Lindley v. Lindley,* 67 N.M. 439, 356 P.2d 455 (1960); *Lynch v. Lictenthaler,* 85 Cal. App.2d 437, 441 (1948); Annot., 169 A.L.R. 9, 65.

23. *See* 7 Wigmore on Evidence § 2106 (3d ed., 1940).

24. *See, e.g., Philippine Sugar Estates Development Co. v. Government of Philippine Islands,* 247 U.S. 385 (1918), an action seeking to "reform" the terms of a written contract, on the grounds that the parties thereto made a mutual mistake of fact.

25. Cal. Evid. Code § 662.

Chapter 6

Presumptions

" 'Presumptions' as happily stated by a scholarly counselor, ore tenus, in another case, 'may be looked on as bats of the law, flitting in the twilight, but disappearing in the sunshine of actual facts.' "

—*Beeman v. Puget Sound Traction, Etc., Co.,* 139 P. 1087, 1088 (Wash. 1914), quoting with approval *Paul v. United Rys. Co.,* 134 S.W. 3 (Mo. 1911)

It was suggested earlier that the hearsay rule is perhaps the one rule of evidence most recognized by name among laymen. Few nonlawyers, though, would profess much knowledge of the arcane complexities of the hearsay rule. The evidence rules called "presumptions" may be only slightly less known by name, but in contrast to the hearsay rule are widely assumed to be understood. Generally speaking, though, presumptions are grossly misunderstood. There is, for example, the often-heard presumption that every man is presumed to know the law. It is one thing to recite that presumption, but how does the presumption operate in the courtroom? Does it mean simply that ignorance of the law is not availing as a defense to a criminal charge? In part. But to take another example, what is the effect in a will contest of the presumption that the maker of the will was of sound mind when he signed his will? Specifically, what becomes of that presumption if it is shown that he was, in fact, in Yeats's words, "mad as birds"? These are the questions this chapter will attempt to answer. It will first describe the nature of presumptions and, more importantly, the nature of their classifications, and then examine some of the more common presumptions. For this purpose, stress will be given the California rules, which have the advantage for the reader of incorporating some of each prevailing point of view on the subject of presumptions. Differences between state and federal rules respecting presumptions will be described at the conclusion of the chapter.

A leading writer on the principles of evidence characterizes "presumption" as

> . . . the slipperiest member of the family of legal terms, except its first cousin, "burden of proof". . . . Agreement can be secured to

this extent, however: a presumption is a standardized practice, under
which certain oft-recurring fact groupings are held to call for uniform
treatment whenever they occur, with respect to their effect as proof to
support issues.[1]

The reasons for the creation and use of presumptions in the law include such factors as
probability and judicial economy. If the issue in an action is whether a letter giving notice of
a matter was received, and the burden is upon the sender to prove its receipt, experience tells
us that if the letter was duly posted, it was probably delivered. Thus the presumption that a
letter duly mailed was received was created for its probability, not to mention the saving of
the court's time in hearing additional evidence that would tend to prove the letter's receipt.
(What becomes of the presumption, should it be shown that the letter was actually not received, is discussed below.)

An additional consideration in the creation of presumptions is fairness to the party with
the burden of producing evidence on the issue. One who posts a letter, unless it is returned as
undeliverable, ordinarily has little means of ascertaining whether the letter was in fact received.
If it is his burden to show the letter's receipt, it is only fair that he enjoy this presumption.

In addition, there are reasons of subtle, often unexpressed matters of social and economic policy that have given rise to presumptions. As Professor McCormick has put it, "[a]
classic instance is the presumption of ownership from possession, which tends to favor the
prior possessor and to make for the stability of estates."[2] The stability of land titles, as noted
in the introduction, is an ancient and much-cited policy of the law.

Thousands of words have been spent in refining the definition and role of presumptions
in the law.[3] To understand the modern and generally held concept of presumptions, it is instructive to examine the work of the drafters of the California Evidence Code, which became
law in 1967. California Evidence Code section 600(a) defines a presumption as "an assumption of fact that the law requires to be made from another fact or group of facts found or otherwise established in the action." Section 600(a) also states "a presumption is not evidence."
Furthermore, section 140 of the California code limits "evidence" to matters "offered" as
proof, thus reinforcing the notion that presumptions are not evidence.

A related but distinct concept is that of an "inference." In California Evidence Code section 600(b), an inference is defined as "a deduction of fact that may logically and reasonably
be drawn from another fact or group of facts found or otherwise established in the action."
Thus, this distinction in definition (which California borrowed from Rule 14 of the Uniform
Rules of Evidence) establishes that a presumption is a deduction (or, rather, an induction) that
the law *requires* to be made, while an inference is one that the law *permits*. An inference then
is simply a matter of logic for the fact-finder. This chapter deals solely with presumptions.

---

1. McCormick on Evidence § 308, at 639 (1954).
2. McCormick on Evidence § 309, at 641 (1954), citing *Oklahoma R. Co. v. Guthrie,* 175 Okla. 40, 52
P.2d 18, 23 (1935); *Guyer v. Snyder,* 133 Md. 19, 104 A. 116 (1918); and, 9 Wigmore on Evidence
§ 2515 (3rd ed., 1940).
3. *See, e.g., McBaine,* Burden of Proof; Presumptions, 2 UCLA L.Rev. 13 (1954); Lowe, The Calif.
Evidence Code: Presumptions, 53 Calif.L.Rev. 1439 (1965); McCormick on Evidence § 309, at
639–643 (1954).

A.   THE CALIFORNIA SYSTEM: A MODERN BLENDING OF VIEWS

Just what is the effect of a presumption? On reflection it should be plain that there are several ways a presumption *can* operate. It can serve to foreclose the issue, permitting no evidence to be received that would tend to dispute the presumed fact. (For example, in California, where such "conclusive" presumptions are recognized, one such presumption holds that the facts recited in an instrument are conclusively presumed to be true as between the parties to the instrument.) The prevailing views, however, hold that a presumption operates either to shift the burden of production on an issue or to shift the burden of persuasion. (Presumptions are the only agents that can cause the burden of persuasion to shift.) These are the two principal, conflicting views; in federal courts, for example, when state law does not supply the rule of decision, a presumption affects *only* the burden of production. (The federal rules are discussed in more detail below.) The approach taken in the California Evidence Code is instructive on the subject of presumptions because it borrows heavily from each of these two views, and as well includes a few conclusive presumptions. Section 601 of California's Evidence Code classifies presumptions as "either conclusive or rebuttable. Every rebuttable presumption is in turn either (a) a presumption affecting the burden of producing evidence or (b) a presumption affecting the burden of proof."[4]

When the California Evidence Code was adopted in 1965, the State Legislature and courts over the years had fashioned scores of presumptions that had not been classified as affecting either the burden of producing evidence or the burden of persuasion. The new Evidence Code codified a number of significant presumptions that had been recognized by the state up to that time and assigned them to one of these two categories. The classifying of other previously recognized presumptions is the task of the courts in California in particular cases.

PRESUMPTIONS AFFECTING THE BURDEN OF PRODUCTION

In California, as mentioned above, a *rebuttable* presumption may be classified as one affecting either the burden of production or the burden of persuasion. As the passages set forth below indicate, the reason for such presumptions is merely to advance the policy of expediting the trial of cases, and presumptions are selected for this policy only if they serve no loftier policy (such as the policy of promoting the stability of titles). The California statute provides the following:

> A presumption affecting the burden of producing evidence is a presumption established to implement no public policy other than to facilitate the determination of the particular action in which the presumption is applied.[5]

To advert to Professor McCormick's rationale for presumptions, those affecting the burden of production only are simply designed to dispense with superfluous proof. The drafters of the California Evidence Code adopted this reasoning:

---

4. Cal. Evid. Code § 601.
5. Cal. Evid. Code § 603; for discussion, *see* 53 Calif.L.Rev. 1445 (1965).

Typically, such presumptions [affecting the burden of production] are based on an underlying logical inference. In some cases, the presumed fact is so likely to be true and so little likely to be disputed that the law requires it to be assumed in the absence of contrary evidence. In other cases, evidence of the nonexistence of the presumed fact, if there is any, is so much more readily available to the party against whom the presumption operates that he is not permitted to argue that the presumed fact does not exist unless he is willing to produce such evidence. In still other cases, there may be no direct evidence of the existence or nonexistence of the presumed fact; but, because the case must be decided, the law requires a determination that the presumed fact exists in light of common experience indicating that it usually exists in such cases.[6]

Because a presumption affecting the burden of production is designed to implement no social or economic policy other than the expeditious resolution of cases, its effect is this: Once the preliminary fact (e.g., the mailing of a letter) has been established, the jury (or the judge when there is no jury) is required to assume the presumed fact (in this example, the receipt of the letter), unless and until evidence is introduced which would support a finding of its nonexistence. Such evidence could consist of, for example, the testimony of the addressee of the letter that he never received it. At that point, the trier of fact must determine the existence or nonexistence of the formerly presumed fact (the receipt of the letter) from the evidence *without regard* to the presumption.[7] The Official Comment to California Evidence Code section 604 explains this operation of a presumption affecting the burden of production as follows:

Such a presumption is merely a preliminary assumption in the absence of contrary evidence, i.e., evidence sufficient to sustain a finding of the nonexistence of the presumed fact. If contrary evidence is introduced, the trier of fact must weigh the inferences arising from the facts that gave rise to the presumption against the contrary evidence and resolve the conflict. For example, if a party proves that a letter was mailed, the trier of fact is required to find that the letter was received in the absence of any believable contrary evidence. However, if the adverse party denies receipt, the presumption is gone from the case. The trier of fact must then weigh the denial of receipt against the inference [*not presumption*] of receipt arising from proof of mailing and decide whether or not the letter was received.

If a presumption affecting the burden of producing evidence is relied on, the judge must determine whether there is evidence sufficient to sustain a finding of the nonexistence of the presumed fact. If there is such evidence, the presumption disappears and the judge need say

---

6.   Official Comment to Cal. Evid. Code § 603.
7.   Official Comment to Cal. Evid. Code § 604.

nothing about it in his instructions. If there is not evidence sufficient to sustain a finding of the nonexistence of the presumed fact, the judge should instruct the jury concerning the presumption. If the basic fact from which the presumption arises is established (by the pleadings, by stipulation, by judicial notice, etc.) so that the existence of the basic fact is not a question of fact for the jury, the jury should be instructed that the presumed fact is also established. If the basic fact is a question of fact for the jury, the judge should charge the jury that, if it finds the basic fact, the jury must also find the presumed fact.

A book that purports to have been published by a public entity is universally presumed to have been so published, and in California (and elsewhere) this presumption is classified as one affecting the burden of production. A manual that purports to have been published by the General Land Office will be presumed to have been so published, and the party offering it as evidence can feel secure that it will be accepted as official if his opponent does not seek to show that it is not. If the opponent, however, does introduce evidence that the book was not published by the G.L.O.—that it was falsely printed, for example—the presumption disappears and each party must rely solely on his evidence.

PRESUMPTIONS AFFECTING THE BURDEN OF PERSUASION
A presumption affecting the burden of persuasion (or "proof") is in the words of the California Evidence Code:

> a presumption established to implement some public policy other than [merely] to facilitate the determination of the particular action in which the presumption is applied, such as the policy in favor of establishment of a parent and child relationship, the validity of marriage, the stability of titles to property, or the security of those who entrust themselves or their property to the administration of others.[8]

Again, the Official Comment to California Evidence Code Section 605 helps to explain (a) the nature of a presumption affecting the burden of persuasion and (b) those characteristics of policy which distinguish it from a presumption affecting merely the burden of production:

> Frequently, presumptions affecting the burden of proof [persuasion] are designed to facilitate determination of the action in which they are applied. Superficially, therefore, such presumptions may appear merely to be presumptions affecting the burden of producing evidence. What makes a presumption one affecting the burden of proof is the fact that there is always some further reason of policy for the establishment of the presumption. It is the existence of this further basis in policy that distinguishes a presumption affecting the burden

---

8.   Cal. Evid. Code § 605; *see also* 53 Calif.L.Rev. 1447 (1965).

of proof from a presumption affecting the burden of producing evidence. For example, the presumption of death from seven years' absence (Section 667) exists in part to facilitate the disposition of actions by supplying a rule of thumb to govern certain cases in which there is likely to be no direct evidence of the presumed fact. But the policy in favor of distributing estates, of settling titles, and of permitting life to proceed normally at some time prior to the expiration of the absentee's normal life expectancy (perhaps 30 or 40 years) that underlies the presumption indicates that it should be a presumption affecting the burden of proof.

Frequently, too, a presumption affecting the burden of proof will have an underlying basis in probability and logical inference. For example, the presumption of the validity of a ceremonial marriage may be based in part on the probability that most marriages are valid. . . .[9]

Thus, the furthering of some policy other than the policy favoring expeditious resolutions of lawsuits is the distinctive characteristic of a presumption affecting the burden of persuasion.

What is at least equally important for an understanding of presumptions, however, is the different *effect* of a presumption affecting the burden of persuasion. As provided in California Evidence Code section 606, "[t]he effect of a presumption affecting the burden of proof is to impose upon the party against whom it operates the burden of proof as to the nonexistence of the presumed fact." Thus, unlike a presumption affecting the burden of production, a presumption affecting the burden of persuasion persists even after the party against whom it operates has produced evidence that would tend to rebut it—or, in the formal rubric of the codes, that would "suffice to sustain a finding against him." It does not at that point disappear, as does a presumption affecting the burden of production, but survives as something the adversary must counter, with the burden of proof having shifted to him.

Consider an action brought by a plaintiff to have a person missing for seven years declared dead. A common presumption in many states, and one affecting the burden of persuasion, is that "[a] person not heard from in seven years is presumed to be dead."[10] Once the basic fact that the man has not been "heard from in seven years" has been established, the presumption that he is dead arises, and the burden of proving that he is alive shifts to the defendant (who had no burden of production proof in the first instance). The defendant must then attempt to produce evidence that the missing man had in fact been heard from during the preceding seven years, all in an effort to overcome the presumption, which does not disappear, that the man is deceased. The Official Comment to California Evidence Code section 606 explains further:

In the ordinary case, the party against whom it is invoked will have the burden of proving the nonexistence of the presumed fact by

---

9.   Official Comment to Cal. Evid. Code § 605.

10.   Cal. Evid. Code § 667; *see, e.g., Benjamin v. Dist. Grand Lodge.* 171 Cal. 260, 266, 152 P. 731 (1915); 4 Calif.L.Rev. 148 (1916); 6 S.Cal.L.Rev. 163 (1933); Annot. 75 A.L.R. 630 (1931).

a preponderance of the evidence. Certain presumptions affecting the burden of proof may be overcome only by clear and convincing proof. When such a presumption is relied on, the party against whom the presumption operates will have a heavier burden of proof. . . .

Thus, it is seen that overcoming a presumption affecting the burden of proof may, depending on the case, require (1) proof amounting to a preponderance of the evidence or (2) proof amounting to clear and convincing evidence, as discussed above. As in the case of determining which presumptions affect the burden of persuasion and not merely the burden of production, the legislature has employed policy considerations to determine which presumptions affecting the burden of proof require clear and convincing evidence to be overcome and which require merely a preponderance of the evidence. Thus, there is in most states a presumption that "[t]he owner of the legal title to property is presumed to be the owner of the full beneficial title." In California, for example, this presumption may be rebutted only by clear and convincing proof.[11] On the other hand, the presumption referred to above that a ceremonial marriage is presumed to be valid may be rebutted by a simple preponderance of the evidence.

SOME TYPICAL PRESUMPTIONS

There are a wide variety of presumptions that are of potential concern to the natural-resource expert. The following are some of these presumptions which have long been recognized in many states and which have been codified in California as presumptions affecting merely the burden of *producing evidence:*

- "Money delivered by one to another is presumed to have been due to the latter." Cal. Evid. Code § 631.
- "The payment of earlier rent or installments is presumed from a receipt for later rent or installments." Cal. Evid. Code § 636.
- "The things which a person possesses are presumed to be owned by him." Cal. Evid. Code § 637.
- "A person who exercises acts of ownership over property is presumed to be the owner of it." Cal. Evid. Code 638.
- "A writing is presumed to have been truly dated." Cal. Evid. Code § 640.
- "A letter correctly addressed and properly mailed is presumed to have been received in the ordinary course of mail." Cal. Evid. Code § .641.
- "A trustee or other person, whose duty it was to convey real property to a particular person, is presumed to have actually conveyed to him when such presumption is necessary to perfect title of such person or his successor in interest." Cal. Evid. Code 642.

---

11. Cal. Evid. Code § 662.

- "A deed or will or other writing purporting to create, terminate, or affect an interest in real or personal property is presumed to be authentic if it:
  (a) Is at least 30 years old;
  (b) Is in such condition as to create no suspicion concerning its authenticity;
  (c) Was kept, or if found was found, in a place where such writing, if authentic, would be likely to be kept or found; and
  (d) Has been generally acted upon as authentic by persons having an interest in the matter." Cal. Evid. Code § 643.
- "A book, purporting to be printed or published by public authority, is presumed to have been so printed or published." Cal. Evid. Code § 644.

Some long-recognized presumptions, which have been codified in California as presumptions affecting the burden of *persuasion,* are the following:

- "The owner of the legal title to property is presumed to be the owner of the full beneficial title." Cal. Evid. Code § 662. (As noted above, in California this presumption may be rebutted only by the introduction of "clear and convincing evidence," which would include such things as a valid easement conveyed by the "owner," an installment contract of sale, etc.)
- "A person not heard from in five years is presumed to be dead." Cal. Evid. Code § 667.

In addition, there are a number of other presumptions found in the California Codes that are not restated in its Evidence Code and to which no definitive court decision has been rendered categorizing them as presumptions affecting the burden of production or burden of proof:

- "A written instrument is presumptive evidence of a consideration."[12]
- Transactions between a trustee and his beneficiary during the existence of the trust, or while the influence acquired by the trustee remains, by which obtains any advantage from his beneficiary, are presumed to be entered into by the latter without sufficient consideration, and under undue influence, and hence a breach of the trustee's fiduciary obligation.[13]
- A signature upon a negotiable instrument such as a draft or promissory note is generally presumed genuine or authorized.[14]

---

12. Cal. Civ. Code § 1614; *see Estate of McConnell,* 6 Cal.2d 493, 499, 58 P.2d 639 (1936).
13. Cal. Probate Code § 16004; *see Bradner v. Vasquez,* 43 Cal.2d 147, 151, 272 P.2d 11 (1954); *Rader v. Thrasher,* 57 Cal.2d 244, 250, 368 P.2d 360 (1962).
14. Cal. U. Com. Code § 3307(1)(b).

- A fee simple title is presumed to be intended to pass by a grant of real property. . . .[15]
- If an interest in property is conveyed to a trustee without naming a beneficiary, it is presumed—so far as innocent third parties are concerned—that the grantee holds the property in his individual right, free from any trust.[16]
- Where no limit is fixed, a tenancy in real property is presumed to be month to month, except in the case of lodgings or dwelling houses, when it is for the length of time adopted for estimation of the rent.[17]

Finally, there are a number of presumptions, common to many states, which have not been codified by California but which have been recognized by court decisions. Depending on the state involved, these may affect either the burden of production or that of persuasion. Among them are the following:

- When the signatures of the testator and the attesting witnesses have been proven genuine, a presumption arises that the will was duly executed. This presumption applies not only in cases in which the witnesses are dead, or unable or unavailable to testify or recollect, but also in cases in which they are adverse to the proponents of the will, or corrupt. Thus this presumption may justify admission of a will to probate despite the testimony of the attesting witnesses of lack of compliance with statutory formalities.[18]
- A will last seen in the possession of the testator, and not found after his death, is presumed to have been revoked by him.[19]
- The recordation of a deed, together with its handing over to another, raises a strong presumption that it was properly delivered. Similarly, possession of the deed by the grantee raises the same presumptions.[20]
- The traditional role of an attorney is such that when he has purported to act on behalf of his client, there is a strong presump-

---

15. Cal. Civ. Code § 1105; *see Basin Oil Co. v. City of Inglewood,* 125 Cal.App.2d 661, 664, 271 P.2d 73 (1954).

16. Cal. Probate Code § 18103; *see Hansen v. G & G Trucking Co.,* 236 Cal.App.2d 481, 490, 46 Cal. Rptr. 186 (1965).

17. Cal. Civ. Code §§ 1943, 1944.

18. *See, e.g., Estate of Pitcairn,* 6 Cal.2d 730, 733, 59 P.2d 90 (1936); *Estate of Braue,* 45 Cal.App.2d 502, 505, 114 P.2d 386 (1941); 76 A.L.R. 617 (1932); 40 A.L.R.2d 1226 (1955).

19. *See, e.g., Estate of LeSure,* 21 Cal.App.2d 73, 83, 68 P.2d 313 (1937); Annot. 3 A.L.R.2d 952, 957 (1949).

20. *See, e.g., Hitch v. Hitch,* 24 Cal.App.2d 291, 293, 74 P.2d 1098 (1934); *Butler v. Butler,* 188 Cal.App.2d 228, 233, 10 C.R. 382 (1961); 141 A.L.R. 308 (1942); *Estate of Galvin,* 114 Cal.App.2d 354, 360, 250 P.2d 333 (1952); 124 A.L.R. 462 (1940).

tion that his acts were within the scope of his employment and
authority.[21]

CONFLICTING PRESUMPTIONS

Occasionally two inconsistent presumptions are invoked by adversaries in a lawsuit for
the purpose of the same issue. These situations are legal oddities that have been resolved by
the courts in a variety of manners. In *City of Montpelier v. Town of Calais,* 114 Vt. 5, 39 A.2d
350 (1944), for example, each party invoked the presumption of official regularity in respect
to the acts of its own officers. In that case the court simply seized the dilemma's horns and
held that the case would be determined without regard to the presumptions.

A different approach, and one that obviously could not be employed where each party is
invoking the same presumption, is found in Rule 15 of the Uniform Rules of Evidence:

> If two presumptions arise which are conflicting with each other
> the judge shall apply the presumption which is founded on the
> weightier considerations of policy and logic. If there is no such pre-
> ponderance both presumptions shall be disregarded.[22]

The California Evidence Code, for all its thoroughness, is silent on the subject of conflicting presumptions. A California court, however, has suggested yet another approach. In
Rader v. Thrasher 57 Cal.2d 244, 252, 368 P.2d 360 (1962), the Court held that the presumption that consideration was lacking when a fiduciary obtains an advantage prevails over the
more general presumption that consideration was present in a transaction memorialized by a
written instrument. The court in that case relied on the principle of statutory construction that
a particular statute prevails over a general statute.

CONCLUSIVE PRESUMPTIONS

The presumptions discussed above, those affecting either the burden of production or
the burden of proof, constitute a larger classification of presumptions called "rebuttable presumptions." In California and several other jurisdictions, a separate but small category of
"conclusive presumptions" is recognized. Conclusive presumptions, again, are those that
simply establish something as a given fact regardless of possibly conflicting evidence. In effect, such presumptions are really substantive rules of law that the parties cannot contradict
by introducing evidence. Among these rules are the following:

- "The facts recited in a written instrument are conclusively presumed
  to be true as between the parties thereto, or their successors in interest; but this rule does not apply to the recital of a consideration."[23]
- "Whenever a party has, by his own statement or conduct, intentionally and deliberately led another to believe a particular thing

---

21. *Gagnon Co. v. Nevada Desert Inn,* 45 Cal.2d 448, 460, 289 P.2d 466 (1955); *Pac. Tel. & Tel. Col. v. Fink,* 141 Cal.App.2d 332, 335, 296 P. 2d 843 (1956).
22. *See also* 44 Harv.L.Rev. 932 (1931); McCormick on Evidence § 312, at 652 (1954).
23. Cal. Evid. Code § 622.

true and to act upon such belief, he is not, in any litigation arising out of such statement or conduct, permitted to contradict it."[24] This is in fact the long-established rule of "equitable estoppel."

- "A tenant is not permitted to deny the title of his landlord at the time of the commencement of the relation."[25]

B.   THE APPROACH OF THE FEDERAL RULES—AND OF 39 STATES

The approach taken by the Federal Rules of Evidence with respect to presumptions is the same as the federal approach to privileges. Federal courts will look to state law for the existence and effect of a presumption when it is invoked by a party on an issue which is controlled by state law. Federal Rule of Evidence 302 provides the following:

> In civil actions and proceedings, the effect of a presumption respecting a fact which is an element of a claim or defense as to which State law supplies the rule of decision is determined in accordance with State law.

For the reader who wishes to probe more deeply into the meaning of this densely stated rule, the remarks of the Advisory Committee on the rules when they were proposed are helpful.[26]

Thus, in a case being tried in federal court in which California law supplies the "rule of decision," presumptions may be invoked that affect not only the burden of production, but the burden of persuasion as well. In all cases, however, where federal and not state law provides the rule of decision, the effect of any presumption is merely to shift the burden of

---

24. Cal. Evid. Code § 623.

25. Cal. Evid. Code § 624.

26. "A series of Supreme Court decisions in diversity cases leaves no doubt of the relevance of *Erie Railroad Co. v. Tompkins,* 304 U.S. 64, 58 S.Ct. 817, 82 L.Ed. 1188 (1938), to questions of burden of proof. These decisions are *Cities Service Oil Col. v. Dunlap,* 308 U.S. 208, 60 S.Ct. 201, 84 L.Ed. 196 (1939), *Palmer v. Hoffman,* 318 U.S. 109, 63 S.Ct. 477, 87 L.Ed. 645 (1943), and *Dick v. New York Life Ins. Co.,* 259 U.S. 437, 79 S.Ct. 921, 3 L.Ed.2d 935 (1959). They involved burden of proof, respectively, as to status as bona fide purchasers, contributory negligence, and nonaccidental death (suicide) of an insured. In each instance the state rule was held to be applicable. It does not follow, however, that all presumptions in diversity cases are governed by state law. In each case cited, the burden of proof question had to do with a substantive element of the claim or defense. Application of the state law is called for only when the presumption operates upon such an element. Accordingly, the rule does not apply state law when the presumption operates upon a lesser aspect of the case, that is, "tactical" presumptions."

"The situations in which the state law is applied have been tagged for convenience in the preceding discussion as "diversity cases." The designation is not a completely accurate one because *Erie* applies to any claim or issue having its source in state law, regardless of the basis of federal jursidiction, and does not apply to a federal claim or issue, even though jurisdiction is based on diversity. Vestal, *Erie R.R. v. Tompkins:* A Projection, 48 Iowa L.Rev. 248, 257 (1963); Hart and Wechsler, The Federal Courts and the Federal System, 697 (1953); 1A Moore, Federal Practice ¶ 0.305[3] (2d ed., 1965); Wright, Federal Courts, 217–218 (1963). Hence the rule employs, as appropriately descriptive, the phrase "as to which state law supplies the rule of decision." *See* A.L.I. Study of the Division of Jurisdiction Between State and Federal Courts, § 2344(c), p. 40, P.F.D. No. 1 (1965)."

going forward with the evidence and not the burden of persuasion. Rule 301 of the Federal Rules of Evidence provides the following:

> In all civil actions and proceedings not otherwise provided for by Act of Congress or by these rules, a presumption imposes on the party against whom it is directed the burden of going forward with evidence to rebut or meet the presumption, but does not shift to such party the burden of proof in the sense of the risk of nonpersuasion, which remains throughout the trial upon the party on whom it was originally cast.

Specific presumptions that have been recognized in federal cases include most of those previously discussed, but include in addition several other interesting presumptions. The Sixth Circuit Court of Appeals, for example, has held that there is a presumption that the law has been obeyed.[27] There is also the generally familiar and sometimes useful presumption that persons are presumed to know the law.[28] Indeed, one Sixth Circuit decision applied a presumption (thought to be wildly optimistic in some circles) that a state's supreme court is familiar with its own prior decisions.[29] In yet another case,[30] the Sixth Circuit held that businessmen were presumed to be aware of dangerous political conditions in the Suez Canal region. Other federal courts have maintained the existence of a presumption that may come as a shock to some—that is, that a government agency never acts unreasonably.[31] In the same vein, the United States Supreme Court has held that administrative agencies are presumed to act according to the law.[32]

There is also a curious presumption that, in the absence of a contrary showing, the law of a foreign country is presumed to be the same as our domestic law. In *Medina v. Hartman,*[33] two Spanish sailors, being detained by American naval authorities, were about to be delivered up to the Spanish authorities for desertion from their ship in San Diego Harbor when they brought a petition for a writ of habeas corpus. No showing was made whether their acts constituted desertion under Spanish law, and, left with that void, the court resorted to the presumption that it was the same as American law.[34] Finally, one other interesting federal presumption is that a party responsible for altering records that have been introduced into evidence did so with the intent to falsify them.[35]

---

27. *TRW, Inc. TRW Michigan Division v. NLRB,* 393 F.2d 771 (6th Cir. 1968).
28. *See, e.g., United States v. Yazell,* 382 U.S. 341 (1966) (a state's law of coverture); *Cole v. Railroad Retirement Board,* 289 F.2d 65 (8th Cir. 1961) (knowledge of United States statutes).
29. *Charney v. Thomas,* 372 F.2d 97 (6th Cir. 1967).
30. *Transatlantic Financing Corp. v. United States,* 363 F.2d 312 (D.C. Cir. 1966).
31. *United States v. Goodloe,* 228 F.Supp. 164 (D.D.C. 1964).
32. *FCC v. Schreiber,* 381 U.S. 279 (1965).
33. 260 F.2d 569 (9th Cir. 1958).
34. *Id.* at 570, footnote 1.
35. *United States, For C.H. Benton, Inc. v. Roelof Const. Co.,* 418 F.2d 1328, 1331 (9th Cir. 1969).

Chapter 7

Privileges: Must the President's Wife Testify Also?

"We are not dealing with one of the vague, undefinable, admonitory divisions of the Constitution whose scope is inevitably addressed to changing circumstances. The privilege against self-incrimination is a specific provision of which it is peculiarly true that 'a page of history is worth a volume of logic.'"

—Justice Felix Frankfurter in *Ullmann v. United States,* 350 U.S. 422, 438 (1956)

The rules of privilege, like the rules excluding hearsay and irrelevant evidence, operate to exclude evidence from introduction in the courtroom. Unlike the rules of relevance and hearsay, however, these rules were not developed for the purpose of assuring the admission of only probative and reliable evidence. (On the contrary, evidence excluded because of a privilege often constitutes the most trustworthy and probative evidence on its point.) Rather, the rules of privilege were shaped to protect certain confidential relationships from breach. In general, it may be said that the law deems the need for confidentiality in certain relationships—marriage, for one—more important than the need for probative evidence at trial.

Several privileges were established at common law, among them the privilege against disclosing confidential communications between an attorney and client. Many of the privileges now commonly recognized in the United States, however, were not recognized at common law. A priest's privilege against disclosing what he has heard in the confessional, for example, was not one that was explicitly recognized at common law.[1]

Where a privilege not to disclose information exists, it applies to all phases of litigation. A client, for example, who exercises his privilege not to disclose confidential advice from his

---

1. 8 Wigmore on Evidence §§2394, 2396 (4th ed., 1974). For the bases of privileges, and a collection of views of the policy reasons for creating them, *see* 8 Wigmore on Evidence, Sections 2192, 2197, and 2285 (4th ed., 1974).

lawyer cannot be compelled to divulge the information while he is testifying at trial, nor in response to interrogatories, nor during his deposition. (*See* Chapter 11.)

It is useful at this point to characterize some of the specific privileges, such as the work-product and attorney–client privilege, that are common to most American jurisdictions today, and to describe the manner in which the privileges are invoked. Where there are substantial differences in the way the state and federal courts view the privileges, these differences will be mentioned.

A.   THE ATTORNEY–CLIENT PRIVILEGE

This privilege, long recognized by the common law, is commonly interpreted by laymen too broadly, chiefly by the failure to distinguish between a client and a consultant. Because its component elements and its way of being invoked parallel several other privileges, it is helpful to begin with this privilege. The privilege in brief attaches to confidential communications between a lawyer and his client. It is necessary to examine each element of that short definition. California Evidence Code section 952 contains a typical definition of a "confidential communication between client and lawyer":

> [I]nformation transmitted between a client and his or her lawyer in the course of that relationship and in confidence by a means which, so far as the client is aware, discloses the information to no third persons other than those who are present to further the interest of the client in the consultation or those to whom disclosure is reasonably necessary for the transmission of the information or the accomplishment of the purpose for which the lawyer is consulted, and includes a legal opinion formed and the advice given by the lawyer in the course of that relationship.

Thus, information conveyed by the client to his attorney in the presence of opposing parties to a negotiation is not privileged, nor is any advice given by the lawyer in the presence of such persons. On the other hand, the presence of the attorney's secretary or clerk, who may be there for the purpose of assisting the lawyer while such communications are made, does not vitiate the privilege.

Because of the difficulty of having to prove affirmatively that a communication was made in confidence, most jurisdictions indulge in a presumption that a communication made between an attorney and client was a confidential communication.[2] It is then incumbent upon the party who would compel disclosure to establish that no confidentiality attached to the communication.

But it should not be thought—and this is a somewhat subtle point—that simply because a client communicates certain information to his attorney in confidence and in the course of the attorney–client relationship, the client is thereby immunized against having to testify

---

2.   *See, e.g.,* Cal. Evid. Code § 917.

about that information. A person's knowledge of certain facts, which is not protected by privilege, does not subsequently become so simply by being communicated to a lawyer. A land surveyor who performs certain operations in the course of conducting a survey and who is later sued for negligence should, to ensure the best possible representation from his attorney, explain thoroughly his actions in conducting the survey. But by so disclosing these matters to his attorney, he is not thereby privileged from testifying about them when called as a witness. He must testify to his operations in the field and in his office, because in those respects he is merely testifying to matters to which he has firsthand knowledge. He may not be compelled to testify as to what he *told* his lawyer—because that is protected by the attorney–client privilege—but assuming that he told his lawyer the truth, it becomes simply a matter of the form of the question. The examiner may properly ask him, "Mr. Whiteman, please explain how you determined that this is the true location of the quarter-corner." He may not ask, however, "Mr. Whiteman, please tell the court what you told your lawyer about your determination of the true location of the quarter-corner." If the witness is truthful—both to his lawyer and to the court—the answer to both questions will be the same, of course.

On rare occasions, the "attorney" may not in fact be authorized to practice law, but if the client reasonably believes him to be, the privilege may nevertheless attach to communications between the two.[3]

A "client" who may enjoy the attorney–client privilege is not confined to "natural persons" (as the law calls people). A corporation, partnership, association, or public entity may all be clients for the purpose of this privilege.

The concept of the "privilege-holder" is important in understanding the nature of a legal privilege. The holder of the attorney–client privilege, for example, is the client, not the lawyer. If the client asserts his privilege, the lawyer may not breach the confidence and disclose the communication unless his services were sought or obtained to enable or aid someone to commit or plan a crime or fraud.[4] Conversely, if the client voluntarily waives his privilege, the attorney no longer has ground to refuse to disclose the matter in question.

The attorney–client privilege may be lost in several ways, most patently by its waiver. As provided in California Evidence Code section 912, the privilege is waived "if any holder of the privilege, without coercion, has disclosed a significant part of the communication or has consented to such disclosure made by anyone." Thus, the inadvertent disclosure of a written legal opinion to one not a party to the attorney–client relationship may not necessarily result in a waiver of the privilege, and the client may assert it in a subsequent legal proceeding. Also, death of the client ordinarily does not mean the dissipation of the privilege; the attorney is still required to claim it on behalf of the decedent. The Vincent Foster case, discussed below, established that principle for the federal courts. There are, however, several instances when the privilege is terminated. The privilege does not attach, for example, to a communication that would tend to show the intention of a deceased client as to a deed, a will, or other instrument purporting to affect an interest in property. For the land expert confronted with an ambiguity in a deed or will of a deceased person then, if he believes the deceased's attorney may have been told things that would resolve the ambiguity, he might bear

---

3. *See, e.g.,* Cal. Evid. Code § 950.
4. For the crime or fraud exception, *see, e.g.,* Cal. Evid. Code § 956.

in mind that the attorney may be compelled to testify.[5] Similarly, there is no privilege for such communications when they are relevant to an issue between people who claim property through a deceased client. In addition, in cases of alleged malpractice of the attorney, the privilege ceases to exist, because otherwise the very evidence needed to prove or disprove the case may be unavailable.

The Federal Rules of Evidence have approached the subject of privileges in a far different manner from their treatment of other rules of evidence, such as the rule against hearsay. When a claim or defense in a lawsuit in federal court is, according to the rules respecting "choice of law," to be decided according to the law of a state, and not according to federal law, privileges are to be determined according to the law of that state. On the other hand, in "federal question" cases (i.e., where the claim or defense is decided according to federal law), privileges "shall be governed by the principles of the common law as they may be interpreted by the courts of the United States in the light of reason and experience."[6] The attorney–client privilege as it has been interpreted by federal courts "in the light of reason and experience" is similar in all significant respects with the characteristics of the privilege discussed above. Thus, the party claiming the attorney–client privilege in the federal courts has the burden of establishing it.[7]

The privilege may extend to agents or other consultants to the attorney required to aid the attorney in counseling his client.[8] Corporations may be deemed clients for purposes of the federal attorney–client privilege, although many difficult questions have arisen about which employees of the corporation are authorized to communicate with an attorney on behalf of the corporation, such that the privilege attaches to those communications.[9]

While the attorney–client privilege in states such as California is absolute unless an exception is specifically stated (such as the crime or fraud exception), a more flexible approach is taken in the federal courts. Certain considerations have been held so strong as to warrant abandonment of the attorney–client privilege by the court.[10]

The federal courts, as much as the state courts, have grappled with the problem of the presence of third parties when a sensitive communication is made between an attorney and his

---

5. *See, e.g.,* Cal. Evid. Code § 960.

6. Fed. R. Evid. 501. The drafters' comment to Rule 501 noted: "there was extreme disagreement with the content of specific rules on privilege at the time these rules were formulated. Consequently the rule was stated generally, with the law being left in its current state, in addition to the requirement that federal courts use state law in civil cases which fall under *Erie R. Co. v. Tompkins,* 304 U.S. 64 (1938)."

7. *United States v. Gurtner,* 474 F.2d 297 (9th Cir. 1973).

8. *United States v. Kovel,* 296 F.2d 918 (2d Cir. 1961) (accountants, ministerial and menial help); *United States v. Brown,* 478 F.2d 1038 (7th Cir. 1973) (accountant); *N.L.R.B. v. Harvey,* 349 F.2d 900 (4th Cir. 1965) (detective); *United States v. White,* 617 F.2d 1131 (5th Cir. 1980) (psychiatrist, but not a psychiatrist merely "sounded out" on the possibility of being employed).

9. *See, e.g., Barr Marine Products Co. v. Borg-Warner,* 84 F.R.D. 631 (E.D. Pa 1979) (a "control group" test was applied by the court to determine who was authorized to deal with counsel in this respect. Once the attorney's advice was related to one who was not a member of the group, the privilege was lost); *In re Grand Jury Subpoena,* 599 F.2d 504 (2d Cir. 1979); *Upjohn Co. v. United States,* 100 S.Ct. 677 (1981).

10. *See, e.g., State of Illinois v. Harper & Row Publishers,* 50 F.R.D. 37 (N.D. Ill. 1969) (loss of memory held to be sufficient reason to obtain a grand jury debriefing memoranda); *Garner v. Wofinbarger,* 430 F.2d 1093 (5th Cir. 1970) (shareholder's suit against corporate management); *Donovan v. Fitzsimmons,* 90 F.R.D. 583 (N.D. Ill. 1981) (alleged pension fund abuse by union officials.)

client. In *United States v. Simpson,* 475 F.2d 934 (D.C. Cir. 1973), the federal court of appeals held that there is no privilege attached to communications between the attorney and client when they had agreed to the presence of a third party who was not an agent of the attorney or of the client. On the other hand, in *United States v. Bigos,* 459 F.2d 639 (1st Cir. 1972), another federal court of appeals held a communication privileged notwithstanding the presence of a third party where the attorney and client clearly intended the communication to be confidential.

Another species of problem arises when there is more than one client, that is, more than one "holder" of the privilege. In *United States v. McPartlin,* 595 F.2d 1321 (7th Cir. 1979), a defendant's statement to the investigator hired by his co-defendant's attorney was held protected by the privilege because it was made in regard to a matter jointly pursued by both defendants. In California, as in many other states, when two or more persons are joint holders of a privilege, such as that for attorney–client communications, a waiver of the privilege by one joint holder does not affect the right of another to claim the privilege.[11]

It should be plain that an expert who is consulted by an attorney does not thereby have an attorney–client relationship such that their communications are privileged. The "client" in such a situation is the person on whose behalf the attorney consulted the expert, not the expert himself. These communications may nonetheless be protected by the attorney–client privilege, notwithstanding that the expert is not the client of the attorney. If the true client has conveyed certain information to his attorney which meets the requirements of the privilege, and the attorney in turn discloses this information to the expert, and if that disclosure "is reasonably necessary for the . . . accomplishment of the purpose for which the lawyer is consulted," many jurisdictions hold that it is protected every bit as much as the original communication from the client to the attorney.[12] Thus, if a client has consulted an attorney with respect to a property-boundary problem, and the attorney requires the advice or services of a land surveyor to properly counsel his client, such a communication would presumably be privileged. It should be pointed out that such communications between an attorney and an expert are generally also protected by the attorney's "work-product" privilege, discussed below. One salient characteristic distinguishing the work-product privilege from the attorney– client privilege, it should be noted, is that the holder of the work-product privilege is generally the attorney, not the client.

In the spring of 1998 the Supreme Court of the United States addressed the question whether the attorney–client privilege—the confidentiality of communications between an attorney and his client—survives the death of the client. The question arose in the course of the meandering investigations of Independent Counsel Kenneth Starr into certain activities of President Bill Clinton, which at first were confined to a 20-year-old real-estate deal. (The scope of Starr's investigation was continually broadened to include, among other things, the President's alleged sexual relationship with a young White House intern, Monica Lewinsky.) A few days before his death, which was reported as a suicide, White House Counsel Vincent W. Foster, Jr. consulted James Hamilton, a prominent Washington lawyer. Hamilton took three pages of handwritten notes during the course of a two-hour interview. Following Foster's death, Starr obtained a subpoena for Hamilton's notes, which Hamilton resisted. Thus the Independent Counsel raised the question whether the attorney–client privilege survives the client's death.

---

11. Cal. Evid. Code § 912.
12. *See, e.g.,* Cal. Evid. Code § 952.

In this politically charged case, the Supreme Court held that the attorney–client privilege survives the client's death. The Court wrote:

> It has been generally, if not universally, accepted, for well over a century, that the attorney–client privilege survives the death of the client in a case such as this. While the arguments against the survival of the privilege are by no means frivolous, they are based in large part on speculation—thoughtful speculation, but speculation nonetheless—as to whether posthumous termination of the privilege would diminish a client's willingness to confide in an attorney. In an area where empirical information would be useful, it is scant and inconclusive.
>
> Rule 501's direction to look to "the principles of the common law as they may be interpreted by the courts of the United States in the light of reason and experience" does not mandate that a rule, once established, should endure for all time. But here the Independent Counsel has simply not made a sufficient showing to overturn the common law rule embodied in the prevailing case law. Interpreted in the light of reason and experience, that body of law requires that the attorney–client privilege prevent disclosure of the notes at issue in this case.[13]

The proponents of the privilege's survival had ready arguments. Clients might well lie to their attorneys if they feared the truth could be bared after their deaths. On the other hand, the adversaries of the privilege's survival had potent arguments of their own. They pointed to a handful of cases that illustrate terrible injustices that were done when lawyers were forbidden to divulge what their dead clients had told them. In 1976, an Arizona judge had invoked the state's law regarding the attorney–client privilege to prevent two lawyers from testifying that a deceased client had confessed to a double murder for which another man was on trial. In Boston's 1990 Charles Stuart case, the Massachusetts Supreme Judicial Court forbade prosecutors to question the lawyer for a man who had first accused an unidentified—and nonexistent—black gunman of killing his wife and who then, as suspicion shifted to himself, jumped to his death a day after consulting the lawyer. The Massachusetts court cited one of its own precedents, from 1833, insisting that "the mouth of the attorney shall be forever sealed."

B.   THE PRIVILEGES FOR AN ATTORNEY'S WORK PRODUCT:
AN ABSOLUTE AND A QUALIFIED PRIVILEGE

The privilege for an attorney's work product, longer recognized by the English courts[14] than American, is too often confused with the attorney–client privilege. One helpful way for the expert to keep the two discretely in mind is to recall that it is this privilege, and not the attorney–client privilege, which *may* afford his consultations to a lawyer privileged against

---

13.   *Swidler & Berlin and James Hamilton v. United States,* 524 U.S. __, 141 L.Ed.2d 379 (1998).
14.   *See, e.g.,* 8 Wigmore on Evidence § 2319 at 618–622, (3d ed., 1940).

disclosure. Also, as mentioned above, another distinguishing characteristic of the privilege is that its holder is the lawyer, not the client. But the point that ought to be impressed is the reason and nature of this privilege: to protect from disclosure the attorney's efforts in preparing his cases so that his client has the benefit of thorough, considered preparation. Imagine the furtive doings of a lawyer who feared he could be deposed to tell all of the weaknesses in his case. Perhaps this large subject can be adequately introduced by noting that there is both an absolute and a conditional work-product privilege.

Some privileges are absolute; in these cases, not even a strong showing of the need for the privileged evidence will dissipate the privilege. One such absolute privilege is that part of the attorney's "work-product" privilege that protects an attorney's research, thoughts, considerations, and notes respecting trial strategy from disclosure; these matters are privileged from disclosure, when the privilege applies, without limitation.

Other matters are only qualifiedly protected from disclosure, however, such as that other portion of the work-product privilege which protects the identities of expert consultants who *may* be called to the witness stand during trial. Just as the work-product privilege protects an attorney's notes and thoughts about trial strategy, it also protects him from being required to disclose the identities of experts he has consulted in the preparation of his case. This latter privilege, however, lasts only so long as the attorney is engaged in the formative preparation of his case. As trial draws near and he decides on what experts he will call, this privilege disappears in favor of another policy of the law: one that encourages full discovery of each side's evidence and testimony before trial, to avoid the generally condemned "trial by surprise."[15] Typically, when the time for trial draws near, each side—through agreement, court order, or court rule—is compelled to identify what expert witnesses it intends to call. This affords the opposing party the opportunity to investigate the backgrounds of these experts and to take their depositions. In this way each side can be prepared to meet the contentions of the other. The privilege still attaches, though, to the identities of experts an attorney has consulted, but whom he has decided not to call as witnesses. (Not surprisingly, a frequent cause for this occurrence is the expert's unfavorable opinion. In major cases he is in this instance often thanked, paid, and provided accommodations in Bermuda for the trial's duration, for added measure to the protection of the work-product privilege.)

The underlying philosophy which has led to both qualified and absolute privileges was expressed very succinctly by Judge Learned Hand in *McMann v. Securities and Exchange Commission,* 87 F.2d 377, 378 (1937):

> The suppression of truth is a grievous necessity at best, more especially when as here the inquiry concerns the public interest; it can be justified at all only when the opposed private interest is supreme.[16]

---

15. The California rule is codified not in its Evidence Code, but among its discovery statutes. It reads: The work product of an attorney is not discoverable unless the court determines that denial of discovery will unfairly prejudice the party seeking discovery in preparing that party's claim or defense or will result in injustice," and "[a]ny writing that reflects an attorney's impressions, conclusions, opinions, or legal research or theories shall not be discoverable under any circumstances." Calif. Code Civ. Proc. § 2018 (b) and (c).

16. *See also* Donnelly, *The Law of Evidence: Privacy and Disclosure,* 14 La.L.Rev. 361 (1954); Barnshart, *Theory of Testimonial Competency and Privilege,* 4 Ark.L.Rev. 377 (1950).

The work-product privilege finds its first well-focused American statement in a 1947 decision of the United States Supreme Court.[17] Some of the Court's remarks in that case still constitute the best exposition of the privilege, and of the reasons for it. In that case, one party, through discovery, was attempting to secure memoranda written by the opposing attorney about witnesses he had interviewed, as well as statements from the witnesses themselves. The Court agreed that the materials sought did not fall under the protection of the attorney–client privilege, but went on to write the following:

> We are . . . dealing with an attempt to secure the production of
> written statements and mental impressions contained in the files and
> the mind of the attorney Fortenbaugh without any showing of necessity or any indication or claim that denial of such production would
> unduly prejudice the preparation of petitioner's case or cause him any
> hardship or injustice. [I]t falls outside the arena of discovery and contravenes the public policy underlying the orderly prosecution and defense of legal claims. Not even the most liberal of discovery theories
> can justify unwarranted inquiries into the files and the mental impressions of an attorney.

Historically, a lawyer is an officer of the court and is bound to work for the advancement of justice while faithfully protecting the rightful interests of his clients. In performing his various duties, however, it is essential that a lawyer work with a certain degree of privacy, free from unnecessary intrusion by opposing parties and their counsel. Proper preparation of a client's case demands that he assemble information, sift what he considers to be the relevant from the irrelevant facts, prepare his legal theories, and plan his strategy without undue and needless interference. That is the historical and the necessary way in which lawyers act within the framework of our system of jurisprudence to promote justice and to protect their clients' interests. This work is reflected, of course, in interviews, statements, memoranda, correspondence, briefs, mental impressions, personal beliefs, and countless other tangible and intangible ways—aptly though roughly termed by the Circuit Court of Appeals in this case as the "work product of the lawyer." Were such materials open to opposing counsel on mere demand, much of what is now put down in writing would remain unwritten. An attorney's thoughts, heretofore inviolate, would not be his own. Inefficiency, unfairness, and sharp practices would inevitably develop in the giving of legal advice and in the preparation of cases for trial. The effect on the legal profession would be demoralizing. And the interests of the clients and the cause of justice would be poorly served.

> We do not mean to say that all written materials obtained or prepared by an adversary's counsel with an eye toward litigation are necessarily free from discovery in all cases. Where relevant and non-privileged facts remain hidden in an attorney's file and where
> production of those facts is essential to the preparation of one's case,
> discovery may properly be had. . . . But the general policy against

---

17. *Hickman v. Taylor*, 329 U.S. 495 (1947).

invading the privacy of an attorney's course of preparation is so well recognized and so essential to an orderly working of our system of legal procedure that a burden rests on the one who would invade that privacy to establish adequate reasons to justify production through a subpoena or court order.

But as to oral statements made by witnesses to Fortenbaugh, whether presently in the form of his mental impressions or memoranda, we do not believe that any showing of necessity can be made under the circumstances of this case so as to justify production.

Denial of production of this nature does not mean that any material, non-privileged facts can be hidden from the petitioner in this case. He need not be unduly hindered in the preparation of his case, in the discovery of facts or in his anticipation of his opponents' position. Searching interrogatories directed to Fortenbaugh and the tug owners, production of written documents and statements upon a proper showing and direct interviews with the witnesses themselves all serve to reveal the facts in Fortenbaugh's possession to the fullest possible extent consistent with public policy. If there should be a rare situation justifying production of these matters, petitioner's case is not of that type.[18]

The "burden" spoken of by the Court, on the party who seeks the work product of an attorney, is an onerous one. It may be met only with a showing of "good cause"; in the case of the identity of experts who an attorney has decided to call as witnesses (as opposed to keeping them in their role as advisers only, or keeping them out of town), that burden is met. In very few other circumstances can a strong enough showing of "cause" be made to permit discovery of an attorney's work product.

C. THE PRIVILEGE AGAINST SELF-INCRIMINATION.

It has long been established in American constitutional law that a defendant in a criminal case has a privilege (a) not to be called as a witness and (b) not to testify. These are not the same privileges. Were this privilege simply one not to testify, a moment's reflection should show the devastating effect of the prosecutor's calling the defendant to the stand, forcing him to exercise his privilege. For this reason, the defendant is free even from being called to the stand, except voluntarily. On the same basis, a person has a privilege to refuse to disclose any matter, whether in a criminal case against him or in some other proceeding, which may tend "to incriminate him"—that is, to show his commission of a crime. These privileges are secured by the Fifth Amendment to the United States Constitution and are frequently found codified in the statutes of the various states, such as in California Evidence Code Sections 930 and 940. The privilege against self-incrimination applies to all manner of crimes, such as perjured expert testimony, fraudulent land surveying, and criminal contempt of court. Because this subject cannot be of any interest to the reader, this chapter shall not tarry on it.

---

18. *Id.* at 508–513.

D.   OTHER PRIVILEGES

There are a number of other privileges, some created by statute and some by court decisions, that arise out of family or professional relationships. These will be discussed briefly.

There are two fundamental privileges recognized in most jurisdictions that arise from the marriage relationship. First, there is the privilege of a married person not to testify against his spouse in a legal proceeding; second, there is a privilege, similar in its application to the attorney–client privilege, to prevent disclosure of confidential communications made between a husband and a wife. As in the case of the attorney–client privilege, this privilege stems from a policy of the law, which in this instance is to foster family harmony by reducing the situations in which marriage partners can be made adversaries in court, particularly against their will.[19] So presumably the President's wife will not be compelled to testify—unless they talked about committing a crime or fraud.

The first spousal privilege, that not to testify against one's spouse in a legal proceeding, is held by the party being *asked* to testify, and not by the other spouse.[20] It is subject to exceptions, the reasons for which are obvious on reflection. For example, the privilege does not apply in a legal proceeding brought by one spouse against the other, a proceeding brought by one spouse to test the mental competence or incompetence of himself or the other, criminal actions brought against one for bigamy, and so on.[21]

The second marital privilege, that respecting confidential communications, is held by each partner to the marriage. The requirements for this privilege are similar to those for the attorney–client privilege. California Evidence Code section 980, for example, provides the following:

> [A] spouse (or his guardian or conservator when he has a guardian or conservator), whether or not a party, has a privilege during the marital relationship and afterwards to refuse to disclose, and to prevent another from disclosing, a communication if he claims the privilege and the communication was made in confidence between him and the other spouse while they were husband and wife.

The confidentiality, and thus the privilege, may be breached by disclosure to third persons.[22] As in the case of the attorney–client privilege, many jurisdictions provide for specifically enumerated exceptions, such as cases in which the communication was made to aid one to commit or plan to commit a crime or fraud.[23] A dissolution of the marriage does not terminate the privilege for communications made during the course of the marriage, but no subsequent utterances are so protected.[24] The federal courts, in "federal question" cases being

---

19. *See, e.g., United States v. Armstrong,* 476 F.2d 313 (5th Cir. 1973).
20. *Trammel v. United States,* 100 S.Ct. 906 (1980); Cal. Evid. Code § 970.
21. *See, e.g.,* Cal. Evid. Code § 972.
22. *See, e.g., United States v. Burks,* 470 F.2d 432 (D.C. Cir. 1972); *Narten v. Eyman,* 460 F.2d 184 (9th Cir. 1969).
23. *See, e.g.,* Cal. Evid. Code § 981; *United States v. Kahn,* 471 F.2d 191 (7th Cir. 1972).
24. *Pereira v. United States,* 347 U.S. 1 (1954).

bound only by prior federal decisions "in the light of reason and experience," have taken a more flexible position than many state courts. Thus, it has been held by the Fifth Circuit Court of Appeals that when a marriage is moribund, although still legally extant, experience and common sense may dictate that the traditional policy reasons for the privilege do not exist.[25]

In many jurisdictions there is a comparable privilege for confidential communications between a patient and his physician, and where this privilege is recognized the holder of it is the patient or his guardian or conservator.[26] As in the case of the attorney–client privilege, the type of communication privileged is that which is customarily made between patients and physicians and does not include discussions that have no relation to medical treatment. While this privilege is widely recognized in state courts, in federal-question actions several federal courts have refused to recognize it.[27]

Again, in jurisdictions where the privilege is recognized, there are exceptions to it that will appear obvious on reflection. There is no privilege for physician–patient communications, for example, in a lawsuit brought by the patient claiming damages for physical injuries, when the physician had been consulted by the patient ostensibly for the treatment of those injuries. Similarly, no privilege is recognized when the patient has brought an action for malpractice against the physician and when the communications concerning the alleged malfeasance of the doctor are germane to the case, or in criminal cases.[28]

In a similar vein, many jurisdictions recognize a psychotherapist–patient privilege[29] and a clergyman–penitent privilege,[30] as well as a privilege for confidential communications between the victim of a sexual assault and his or her counselor.[31]

In some jurisdictions, a public entity holds a privilege to refuse to disclose "official information," which was acquired in confidence by a public employee in the course of his duty, and which has not been officially disclosed to the public before a claim of privilege is made.[32] In criminal cases, there is also a qualified privilege protecting the identity of an informant.[33]

Privileges have also been recognized for such diverse matters as how one voted at a public election by secret ballot,[34] as well as for trade secrets.[35]

---

25. *United States v. Cameron,* 556 F.2d 752 (5th Cir. 1977); *see also Ryan v. CIR,* 568 F.2d 531 (7th Cir. 1977).

26. *See, e.g.,* Cal. Evid. Code §§ 990–994.

27. *United Staes v. Mancuso,* 444 F.2d 691 (5th Cir. 1971); *cf. Catoe v. United States,* 131 F.2d 16 (D.C. Cir. 1942).

28. *See, e.g.,* Cal. Evid. Code §§ 996–1007.

29. *See, e.g.,* Cal. Evid. Code §§ 1010–1028.

30. *See, e.g.,* Cal. Evid. Code §§ 1030–1034; *United States v. Luther,* 481 F.2d 429 (9th Cir. 1973); *Mullen v. United States,* 263 F.2d 275 (D.C. Cir. 1958); *cf. United States v. Wells,* 446 F.2d 2 (2d Cir. 1971).

31. *See, e.g.,* Cal. Evid. Code §§ 1035–1036.2, *and see* §§ 1037–1037.1.

32. *See, e.g.,* Cal. Evid. Code § 1040. *Re* official secrets privilege *see United States v. Reynolds,* 345 U.S. 1 (1953); *Carr v. Monroe Mfg. Co.,* 431 F.2d 384 (5th Cir. 1970); *Ethyl Corp. v. Environmental Protection Agency,* 478 F.2d 47 (4th Cir. 1973).

33. *See, e.g.,* Cal. Evid. Code § 1041; *Rugendorf v. United States,* 376 U.S. 528 (1964); *Roviaro v. United States,* 353 U.S. 53 (1957). Generally speaking, the accused in a criminal case bears a heavy burden to show that the identity of an informant is necessary to his defense. *United States v. Skeens,* 449 F.2d 1066 (D.C. Cir. 1971); *United States v. Kelly,* 449 F.2d 329 (9th Cir. 1971).

34. Cal. Evid. Code § 1050.

35. Cal. Evid. Code § 1060.

Most states as well as the federal courts do not recognize an accountant–client privilege, nor a banker–customer privilege.[36] Similarly, while some states recognize a privilege of a journalist to protect his sources, such a privilege is not absolute in the federal courts.[37] A privilege has occasionally been recognized for matters that would tend to bring "disgrace" on the person asked to testify; it is rarely recognized any longer.[38]

Finally, there is the issue of whether a witness's claim of a privilege is a matter that can be broached by opposing counsel to the jury. Some jurisdictions permit such argument as, "Ladies and gentlemen of the jury, you heard Mr. Jones invoke a privilege not to testify; let me suggest what his testimony would have been. . . ." Perhaps most jurisdictions today do not.[39]

---

36. *See, e.g., Couch v. United States,* 409 U.S. 322 (1973); *Harris v. United States,* 413 F.2d 316 (9th Cir. 1969).
37. *Branzburg v. Hayes,* 408 U.S. 665 (1972).
38. *See* McCormick on Evidence § 81 at 166 (1954).
39. *Compare* Cal. Evid. Code § 913 *with* McCormick on Evidence § 80 at 163 (1954).

Chapter 8

How to Prove the Earth is Round: The Notion of Judicial Notice

"We cannot as judges be ignorant of that which is common knowledge to all men."

—Justice Felix Frankfurter in *Sherrer v. Sherrer,* 334 U.S. 343, 366 (1948)

Certain matters, it should be evident, are more or less self-evident and ought to require no, or little, formal proof. There are 365 days in a year; February has 28 days except in leap years; an event which occurred on December 25 occurred on Christmas. These facts are said to be "judicially noticed"; their notoriety is such that the court will find them to be so, or direct the jury to, dispensing with the need for proof.

The Bureau of Land Management is within the Department of Interior. Mercury is the planet closest to the sun. Are these propositions "judicially noticeable?" What are the limits of "judicial notice," and what predicates are there for a court's taking notice of certain facts without the usual strict requirements of proof? This subject, while seemingly simple, is nonetheless one of the most elusive subjects of evidence law. If this chapter confounds more than it enlightens, perhaps it does so because the law of judicial notice is more perplexing than illuminating. Also, it is ever-evolving.

A beginning must be found somewhere, and perhaps an older, standard legal dictionary is as good a place as any. The definition of "judicial notice" contained in *Black's Law Dictionary* nearly 50 years ago (4th ed., 1951, p. 986) gives a glimpse of the concept of judicial notice," but, as discussed below, is much more restrictive than the modern notion:

> *Judicial Notice.* The act by which a court, in conducting a trial, or framing its decision, will, of its own motion, and without the production of evidence, recognize the existence and truth of certain facts,

having a bearing on the controversy at bar, which, from their nature, are not properly the subject of testimony, or which are universally regarded as established by common notoriety, e.g., the laws of the state, international law, historical events, the constitution and course of nature, main geographical features, etc. [Citations.] The cognizance of certain facts which judges and jurors may properly take and act upon without proof, because they already know them. [Citation.]

The true conception of what is "judicially known" is that of something which is not, or rather need not be, unless the tribunal wishes it, the subject of either evidence or argument. [Citation.] The limits of "judicial notice" cannot be prescribed with exactness, but notoriety is, generally speaking, the ultimate test of facts sought to be brought within the realm of judicial notice; in general, it covers matters so notorious that a production of evidence would be unnecessary, matters which the judicial function supposes the judge to be acquainted with actually or theoretically, and matters not strictly included under either of such heads. [Citation.]

Some of the respects in which this definition is unduly restrictive today are these: (1) Judicial notice of certain facts may now be obtained upon the motion of a party, and not merely upon the court's own motion; (2) frequently the production of some evidence is required before judicial notice will be invoked; and (3), facts that are (a) susceptible of immediate and accurate determination and (b) not subject to reasonable dispute may be judicially noticed in many jurisdictions, as much as facts that are already notorious.

A digression is in order to relate a story that illustrates the use of judicial notice. An obscure land surveyor of the last century who regressed to take up the practice of law is in this account reported to have made dramatic use of the principle of judicial notice in his first defense in a murder trial. The account is by Judge J. W. Donovan in his book, *Tact in Court:*

Grayson was charged with shooting Lockwood at a camp-meeting, on the evening of August 9, 18—, and with running away from the scene of the killing, which was witnessed by Sovine. The proof was so strong that, even with an excellent previous character, Grayson came very near being lynched on two occasions soon after his indictment for murder.

The mother of the accused, after failing to secure older counsel, finally engaged young Abraham Lincoln, as he was then called, and the trial came on to an early hearing. No objection was made to the jury, and no cross-examination of witnesses, save the last and only important one, who swore that he knew the parties, saw the shot fired by Grayson, saw him run away, and picked up the deceased, who died instantly.

The evidence of guilt and identity was morally certain. The attendance was large, the interest intense. Grayson's mother began to wonder why "Abraham remained silent so long and why he didn't do something!" The people finally rested. The tall lawyer (Lincoln) stood up and eyed the strong witness in silence, without books or notes, and slowly began his defence by these questions:

Lincoln.   "And you were with Lockwood just before and saw the shooting?"

Witness.   "Yes."

Lincoln.   "And you stood very near to them?"

Witness.   "No, about twenty feet away."

Lincoln.   "May it not have been *ten* feet?"

Witness.   "No, it was twenty feet *or more.*"

Lincoln.   "In the open field."

Witness.   "No, in the timber"

Lincoln.   "What kind of timber?"

Witness.   "Beech timber."

Lincoln.   "Leaves on it are rather thick in August?"

Witness.   "Rather."

Lincoln.   "And you think *this* pistol was the one used?"

Witness.   "It looks like it."

Lincoln.   "You could see defendant shoot—see how the barrel hung, and all about it?"

Witness.   "Yes."

Lincoln.   "How near was this to the meeting place?"

Witness.   "Three-quarters of a mile away."

Lincoln.   "Where were the lights?"

Witness.   "Up by the minister's stand."

Lincoln.   "Three-quarters of a mile away?"

Witness.   "Yes,—I answered ye *twiste.*"

Lincoln.   "Did you not see a candle there, with Lockwood or Grayson?"

Witness.   "No! what would we want a candle for?"

Lincoln.   "How, then, did you see the shooting?"

Witness.   "By moonlight!" (defiantly).

Lincoln.   "You saw this shooting at ten at night—in beech timber, three-quarters of a mile from the light—saw the pistol barrel—saw the man fire—saw it twenty feet away—saw it all by moonlight? Saw it nearly a mile from the camp lights?"

Witness.   "Yes, I told you so before."

The interest was now so intense that men leaned forward to catch the smallest syllable. Then the lawyer drew out a blue covered almanac from his side coat pocket—opened it slowly—offered it in evidence—showed it to the jury and the court—read from a page with careful deliberation that the moon on that night was unseen and only arose at *one* the next morning.

Following this climax Mr. Lincoln moved the arrest of the perjured witness as the real murderer, saying: "Nothing but a *motive to clear himself* could have induced him to swear away so falsely the life of one who never did him harm!" With such determined emphasis did Lincoln present his showing that the court ordered Sovine arrested, and under the strain of excitement he broke down and confessed to being the one who fired the fatal shot himself, but denied it was intentional.

The almanac—otherwise wholly inadmissible under the rules of evidence—permitted the jury to take judicial notice of the time of the moon's rise that night.[1]

An 1896 California case, reported in the official judicial reports of that state, contains an account of the very same use of judicial notice. *People v. Mayes*[2] was a prosecution for the theft of a steer. A witness for the defendant testified that he observed someone other than the defendant steal the steer on the night in question "betwixt 9 and 10, I suppose." The judge in his charge to the jury instructed them that as a matter of judicial knowledge the moon on that night rose at 10:57 p.m. The defendant, trying to place the time of moonrise as early as possible, had sought by affidavit to show that the moon on the night in question rose at 10:35 p.m. The California Supreme Court, however, held that the judge, by consulting reliable calendars and almanacs, had properly taken judicial notice of the correct time of the rising of the moon.[3]

This principle of judicial notice, as you have it, certainly seems so appropriate, so self-evidently necessary for the efficient trial of cases, that one could say its limits should be dictated simply by the exigencies of the individual case. Such a rule, the following examples should show, would be perversely unworkable in a system that seeks to be just. A quite com-

---

1.   This account is reprinted in Francis L. Wellman's classic work, *The Art of Cross-Examination*, 55–57 (4th ed., 1936). The reader may be interested in Wellman's remarks following his relating of this story, though they have nothing to do with the subject of this chapter:

> I have quoted this occurrence verbatim as given by Judge Donovan. It affords a most striking illustration of the "fallacies of testimony." The occasion on which Lincoln acquitted his client of a charge of murder by confronting an eye witness with an almanac to refute the testimony given "by the light of the moon," instead of being the first criminal case tried by "young" Abraham Lincoln, was in reality one of the last and most important criminal cases he ever tried. The defendant's name instead of being Grayson was William Armstrong, who was tried August 29, 1857, for the killing of one James Metzker, not Lockwood; and it was upon this occasion that Lincoln's talents as a trial lawyer saved the day for his client.
>
> The story of this now famous case has often been recounted, and the distortions wrought by many versions of it, through many mouths and during many years, might well take a prominent place in the discussion of the *unreliability of honest testimony*. . . . . [Emphasis added.]

Wellman, *supra* at 57. The story is also related by Lincoln's biographer, Carl Sandburg, in his *Life of Lincoln*, Volume 1, pp. 169–170.

2.   113 Cal. 618, 45 P. 860 (1896).

3.   *Id.* at 624–625.

parable principle was invoked by the ecclesiastical jurists who presided at the trial of Galileo; then, too, it was invoked on a matter of astronomy. From The *Times of London*, October 23, 1980, we have the report headlined "Vatican is to Reexamine Galileo's heresy." The short item reads, "The Pope has called for a fresh inquiry 'with full objectivity' into the case of Galileo, the great astronomer and mathematician imprisoned by the Catholic Church in the 17th Century for being 'vehemently suspected of heresy' [for having agreed with Copernicus that the earth revolves around the sun]."[4] The ecclesiastical court had but taken what we would call "judicial notice" that the sun revolves around the earth. A second example is a postscript by Mr. Wellman to the story of Lincoln's first, or last if you please, murder defense. Mr. Wellman concludes the several accounts of the Grayson or Armstrong case with the following remarks:

> The comparison of these two accounts of a very simple and familiar method of cross-examiners serves as a most striking illustration of the "fallibilities of testimony," for the *details* of the Armstrong case have been gossip at the Illinois bar almost to the present day, and the original story as given by Mr. Hill has evidently gradually reached the form in which it is given by Judge Donovan. The *main* feature of the examination was the same—the use of the calendar [or almanac],—but the names of the defendant, and of the witness, and all the details of the occurrence both before and after the trial are entirely different. It has even been frequently stated by members of the Illinois bar that Lincoln played a trick on the jury in this case by substituting an old calendar for the one of the year of the murder and virtually manufactured the testimony which carried the day. This rumor has been repeatedly exposed, but I am told it still persists on the Illinois circuit to this day.[5]

These stories should illustrate then not only the potential usefulness of judicial notice, but also the need for caution in its invoking. The determination of a court to take judicial notice of facts is a matter of some discrimination, entailing distinctions not unlike those found in the formulation of the hearsay rule and its exceptions, but more often overlooked. Before considering this matter of caution in more detail, and considering the kinds of matters of interest to the natural-resource expert which have been traditionally noticed by the courts, it is useful to review the general rules of judicial notice.

In California, which is representative of many states today, matters subject to judicial notice are divided into two categories: (1) matters that a court *must* judicially notice and (2) those that a court *may* judicially notice. Among the matters that must be judicially noticed are (a) the decisional (i.e., reported cases), constitutional, and statutory law of the state and the United States, (b) rules of pleading, practice, and procedure of all the federal courts, (c) "the true signification of all English words and phrases and of all legal expressions," and

4. *The Times of London*, Oct. 23, 1980 at 6, col. 5.
5. Wellman, *supra*, note 1, at 59.

(d) "facts and propositions of generalized knowledge that are so universally known that they cannot reasonably be the subject of dispute."[6]

With respect to the second category, a California court *may* take judicial notice of "facts and propositions that are of such common knowledge within the territorial jurisdiction of the court that they cannot reasonably be the subject of dispute," and "facts and propositions that are not reasonably subject to dispute and are capable of immediate and accurate determination by resort to sources of reasonably indisputable accuracy."[7] The location of streets in the city in which the court sits are matters of "common knowledge within the territorial jurisdiction of the court." The location of streets in Bremen, Ohio, on the other hand, is not a matter of common knowledge in San Francisco, and thus the whereabouts of Elm Street in Bremen could not be judicially noticed by a San Francisco court on that ground. With some creative persuasion, however, perhaps it could be judicially noticed by a San Francisco court on the second, permissive ground of judicial notice—that it is capable of determination through reliable sources, such as published maps. This second category—facts capable of immediate and accurate determination by resort to sources of reasonably indisputable accuracy—comprises such other matters as the date of Galileo's birth. It is supposed that few people would be able to recite that date without consulting, for example, the *Encyclopaedia Britannica.* Yet it would be ludicrous to require formal proof.

Other matters that the California statute permits, but does not require, to be judicially noticed include the law of any other state of the United States, federal regulations, official acts of the legislative, executive, and judicial departments of the federal government and of any state, court records, and foreign law.[8]

The California statute also contains a device enabling any party to *require* the court to judicially notice the types of facts described in the second (discretionary) category. The statute provides that the court *shall* take judicial notice of any matter that *may* be judicially noticed if a party requests it, gives sufficient notice to his adversary, and provides the court with the means to obtain the knowledge (e.g., pertinent passages from an encyclopedia). This is an example of a modern use of judicial notice in which some evidence is actually received, though falling far short of formal proof. In a court-tried case, the judge merely considers the matter he has judicially noticed in the course of arriving at his decision; in a jury trial, the court instructs the jury that it is to consider as fact the matter in question (the layout of city streets, the location of city boundaries, etc.).

The Federal Rules take a considerably different tack. Rule 201 of the Federal Rules of Evidence, which is the sole federal rule dealing with judicial notice, states decisively that it "governs only judicial notice of adjudicative facts." It provides the following in subsection (b): "A judicially noticed fact must be one not subject to reasonable dispute in that it is either (1) generally known within the territorial jurisdiction of the trial court or (2) capable of accurate and ready determination by resort to sources whose accuracy cannot reasonably be questioned."

The federal advisory committee that urged that judicial notice be confined to "adjudicative facts" noted that the law of foreign countries ought not to be a subject of judicial notice,

---

6. Cal. Evid. Code § 451.
7. Cal. Evid. Code § 452(g) and (h).
8. Cal. Evid. Code § 452(a) through (f).

at least in those terms, because it was already governed by Rule 44.1 of the Federal Rules of Civil Procedure. There, the determination of a matter of foreign law is to be treated not as judicial notice of a fact, but in the same manner as the process whereby a court ascertains a matter of domestic law:

> A party who intends to raise an issue concerning the law of a foreign country shall give notice in his pleadings or other reasonable written notice. The court, in determining foreign law, may consider any relevant material or source, including testimony, whether or not submitted by a party or admissible under the Federal Rules of Evidence. The court's determination shall be treated as a ruling on a question of law.

A 1982 decision of the California Supreme Court shows what a profound effect on land title may be worked by judicial notice of the content of foreign law. In *City of Los Angeles v. Venice Peninsula Properties,*[9] the City of Los Angeles and the State of California contended that certain alleged "tidelands" within the boundaries of a Mexican land grant, the Rancho Ballona, were subject to a "public trust" interest held by the State of California. One of the underpinnings of this argument was that these tidelands were so held by the Government of Mexico at the time the grant was made in 1839, prior to the annexation of California by the United States in 1848. The land grant had been made by the Governor of California to four "Californios," Augustine and Ignacio Machado and Philipe and Tomas Talamantes. It was thus imperative that the City and State somehow establish that such public rights existed during Mexico's reign over California. The California Supreme Court, when it addressed this issue, stated simply, "the law of Mexico at the time of cession [1848] declared that the public had a right to the use of the tidelands; this right was similar to the common law public trust." To take judicial notice of this proposition, the California Supreme Court consulted an English translation of *Las Siete Partidas* (the codified Mexican law at the time), the testimony of a witness qualified at trial as an expert in Mexican law, and an article that had recently been published by an Assistant Attorney General for the State.[10] From these three sources, the Court took judicial notice of a foreign law that briefly, but radically, changed the status of many coastal land titles in California—that is, until the United States Supreme Court put things right and reversed the California Supreme Court. *Summa Corp. v. Los Angeles,* 466 U.S. 198, 80 L.Ed.2d 237 (1984).

The purpose of the drafters of the Federal Rules in limiting judicial notice to "adjudicative facts," according to their notes on Rule 201, was to disallow judicial notice of "legislative facts." An example given by the drafters is the case of *Hawkins v. United States,* 358 U.S. 74 (1958), in which the United States Supreme Court refused to discard the common law rule that one spouse could not testify against the other. The Court in that case had remarked that "[a]dverse testimony given in criminal proceedings would, we think, be likely to destroy almost any marriage."[11] The drafters, having noted that the fundamental principle of

---

9. 31 Cal. 3d 288 (1982).

10. *Id.* at 2997, fn. 8 and 9.

11. 358 U.S. at 78.

judicial notice ought to be the indisputability of the proposition, remarked that "this conclusion has a large intermixture of fact, but the factual aspect is scarcely 'indisputable'." Having thus explained their meaning of "legislative facts," the drafters go on to explain the expression adjudicative facts":

> In view of these considerations, the regulation of judicial notice of facts by the present rule extends only to adjudicative facts.
> What, then, are "adjudicative" facts? [Professor] Davis refers to them as those "which relate to the parties," or more fully:
> "When a court or an agency finds facts concerning the immediate parties—who did what, where, when, how, and with what motive or intent—the court or agency is performing an adjudicative function, and the facts are conveniently called adjudicative facts. . . .
> "Stated in other terms, the adjudicative facts are those to which the law is applied in the process of adjudication. They are the facts that normally go to the jury in a jury case. They relate to the parties, their activities, their properties, their businesses."[12]

Under both the California and federal systems, judicial notice of all but the most obvious matters—the date on which Christmas falls, for example—is ordinarily initiated by the request of a party. While such a request may be made orally at a hearing, the requirement that the opposing party be given sufficient notice of the request to meet it with any contrary information ordinarily requires that a request of the court to take judicial notice be made in writing. This may be done in a document styled as a "Request that the Court Take Judicial Notice," or it may be made in the text of a brief or in another paper filed with the court.

With this theoretical background, some specific matters that have been judicially noticed, and that are of importance in land and natural-resource cases, will be reviewed. The reader who has reflected on the rules of evidence thus far will be only too acutely aware of the need for such a principle as judicial notice to establish critical points of a case. The principle is elusive, however, and may be more readily understood from actual examples than from abstract statements.

As a general proposition, courts will take judicial notice of customs or usages that are commonly known, and of commercial usages and customs that have become well-established and thus are part of the "law merchant" of the country.[13] "The general customs and course of business in any profession, trade, or occupation, if they are sufficiently notorious, are proper subjects of judicial notice."[14] In the real-property field, courts have taken judicial notice of many practices in real-estate transactions. One federal court, for example, has taken judicial notice of the fact that a recited consideration of "$10.00" was inserted into the agreement to conceal the true consideration for the transaction and that the $10.00 recited in such agreements seldom, if ever, changes hands.[15] Another court has taken judicial notice of the fact

---

12. Fed. R. Evid. 201, Note of Advisory Committee on Proposed Rules (1982).

13. 29 Am.Jur. 2d, *Evidence* §§ 86–87, at 121 (1967).

14. 31 Cal.Jur.3d, *Evidence* § 68 at 98 (1976).

15. *United States v. Certain Parcels of Land, etc.*, 85 F.Supp. 986, 1006 (S.D. Cal. 1949).

that it is "a very common thing" for a grantee of real property to obtain a policy of title insurance at the time of recordation of his deed.[16]

With respect to the surveys and methods of disposal of the public lands, a broad variety of records of the General Land Office and of the Bureau of Land Management, including plats, field notes, and the like, are subject to judicial notice.[17] "Courts may take judicial notice of government surveys of public land as official acts of the executive department of the United States, or of this State . . . ."[18]

The principle that official government surveys, maps, and other land records are to be judicially noticed has long been the law in England and Canada as well.[19]

The extent to which judicial notice may be taken in these respects, thus eliminating the need for extensive formal proof, is well-illustrated in the following passage from a 1906 decision of the California Supreme Court:

> In disposing of its public lands the government causes them to be laid off into townships six miles square, bounded by meridian lines on the east and west and by parallels of latitude on the north and south. This, at least, is the ideal township, but owing to faulty location of the boundary lines or to proximity to the seacoast, or to both causes, as in this case, and to the convergence of meridian lines—as in all cases—a township never contains thirty-six sections each exactly a mile square, into which they are supposed to be divided. The rules for making the subdivisional surveys, however, are well understood, and there is a uniform method of numbering and designating the sections and their smaller subdivisions, of all of which matters the courts take judicial notice. Section 31 is at the southwest corner of the township, next on the east is section 32, and so on to section 36 at the southeast corner. We also take judicial notice that under the laws of the United States and the regulations of the land office no sales of public lands are made until the plat and field notes of the subdivisional survey of the township in which they lie have been returned and approved, and that the patents issued to purchasers describe the lands patented as they are delineated on the approved plat—which remains a public record in the office of the surveyor-general, and copies of which, officially certified, are deposited in the local land office. We also know that it is made the duty of the surveyors employed to make

---

16. *Mutual B-L Assn. v. Security T.I. & G. Co.,* 14 Cal.App.2d 225, 231 (1936).
17. *Livermore v. Beal,* 18 Cal.App.2d 535, 541 (1937).
18. 31 Cal.Jur.2d, *Evidence* §37, at 73 (1976); *Rogers v. Cady,* 104 Cal. 288, 290–291. (1894); *Harrington v. Goldsmith,* 136 Cal. 168, (1902); *Stein v. Ashby,* 24 Ala. 521. (1854); *White v. State of California,* 21 Cal.App.3d 738, 762 (1971).
19. England: *Evans v. Merthyr Tydfil,* 1 Ch. 241 (1899) (a survey of land belonging to the Crown, made under the provisions of a statute and filed in the Law Revenue Office). Canada: *Badgeley v. Bender,* 3 U.C. Rep. O.S. 221 (1834) (official surveyor map admissible); *O'Connor v. Dunn,* 2 Ont.App. 247 (1877) (surveyor's official survey held admissible); *contra Maynes v. Dolan,* 3 All. N.D. 573 (1857) (Crown survey not admitted to show that the line had actually been run).

> these subdivisional surveys to mark the section corners with posts
> properly lettered, or other monuments, and also the intersection of
> quarter-section lines with the section lines, and more especially to
> preserve accurate field-notes of their surveys showing the natural fea-
> tures of the country, such as streams, mountain ridges, the ocean
> shore, nature of the soil, whether barren or fertile, wooded or meadow;
> distance and bearing of witness-trees from section and quarter-section
> corners, etc.[20]

The California Supreme Court took judicial notice of all of the propositions recited in the passage just quoted. To appreciate the usefulness of judicial notice, consider what testimony and documentary evidence would be required to establish each of these matters if there were no such rule.

Courts have, in addition, taken judicial notice that the segregation surveys of swamp and overflowed lands are the government's official identification of the lands that were granted to the state by the Arkansas Swamp Lands Act of 1850.[21] In another case, the same court refused to hold a deed invalid for its failure to state the meridian from which the range in which the subject property lay was numbered; the court judicially noticed that there was only one meridian for the ranges in the county.[22] Judicial notice has also been taken that, pur-suant to the federal laws and the regulations of the General Land Office, no sales of public lands were to be made until the plat and field notes of the survey of the township had been returned and approved.[23]

In a peculiar application of judicial notice in this field, the California Supreme Court has observed that the inaccuracy of the early surveys in California, as well as in other states, is a matter of such common knowledge that the courts are warranted in taking cognizance of the fact by the vehicle of judicial notice.[24]

Courts have often taken judicial notice of the records of the National Ocean Survey, in-cluding its hydrographic and topographic surveys of the coasts of the United States. (The Na-tional Ocean Service was originally called the "United States Coast Survey," and later the "United States Coast and Geodetic Survey," and more recently the "National Ocean Survey," the "National Ocean Service," and again the "National Ocean Survey.") These surveys are not cadastral or boundary surveys; they were conducted for the primary purpose of produc-ing nautical charts for the mariner. Nevertheless, they are frequently of indispensable utility in boundary cases, especially riparian and littoral boundary cases. While water bodies were meandered in the course of the public land surveys, the meander lines run in the course of such surveys are not boundary lines; they are merely approximations for the purpose of as-certaining the amount of acreage contained in the subdivision. The United States Supreme Court has stated the following:

---

20. *Kimball v. McKee,* 149 Cal. 435, 439–440 (1906).
21. *Foss v. Johnstone,* 158 Cal. 119, 110 P. 294 (1910).
22. *Faekler v. Wright,* 86 Cal. 217, 24 P. 996 (1890).
23. *Saunders v. Polich,* 250 Cal.App.2d 136, 58 Cal.Rptr. 198 (1967).
24. *Hellman v. Los Angeles,* 125 Cal. 383, 387, 58 P. 10 (1899).

Meander lines are run in surveying fractional portions of the public lands bordering upon navigable rivers, not as boundaries of the tract,
but for the purpose of defining the sinuosities of the banks of the
stream, and as the means of ascertaining the quantity of the land in the
fraction subject to sale, and which is to be paid for by the purchaser.

In preparing the official plat from the field notes, the meander
line is represented as the border line of the stream, and shows, to a
demonstration, that the water-course, and not the meander line, as actually run on the land, is the boundary.[25]

In waterfront-boundary cases, then, it is often necessary to determine the true location of the
shoreline at the time of the original government survey, to compare it to the meander line
that was run by the government surveyor. In addition, waterfront boundaries may migrate
with gradual changes in the location of the shoreline. These changes may be caused by accretion, erosion, reliction, or submergence. Accretion is the gradual building up of land by
the deposition of sediments, or "alluvion." Erosion is the gradual wearing away of land. Reliction is the recession of the shoreline by the lowering of the water surface. Submergence is,
well, obvious.[26] (When a sudden shoreline change occurs, an "avulsion," the property boundary remains in its former location.)[27]

Thus, because of the meander-line principle, and also because the boundary of the water
body in question may have changed in location from the time of the original survey, courts in
waterfront-boundary cases have frequently found need for the hydrographic and topographic
surveys done by NOS or its predecessors. A representative statement of the views of the
courts in this regard is contained in the case of *United States v. Romaine,* 255 F. 253,
254–255 (9th Cir. 1919), before the United States Court of Appeals for the Ninth Circuit. In
that case the dispute involved the question of the true location of the mouth of the Nooksack
River, Washington, in 1855.

We are unable to agree with the trial court as to the effect which
should be given to the hydrographic maps of the United States Coast

---

25. *St. P. & P.R.R. Co. v. Schurmeier,* 74 U.S. (7 Wall.) 272, 286–287 (1869). This principle has been re-
iterated on numerous occasions by the courts. *See, e.g., Jeffers v. East Omaha Land Co.,* 134 U.S. 178, 196
(1890); *Hardin v. Jordan,* 140 U.S. 371, 380–381 (1891), and cases there cited; *City of Los Angeles v. Borax
Consolidated Ltd.,* 74 F.2d 901, 902 (1935), *aff'd sub nom. Borax Consolidated v. Los Angeles,* 296 U.S. 10
(1935). Moreover, this principle is set forth, and the passage from *Schurmeir* case quoted, in the government's *1947 Manual of Instructions for the Survey of the Public Lands of the United States,* pages 232–233,
and in its *1973 Manual,* at pp. 93–98. *See also* 2 Shalowitz *Shore and Sea Boundaries* 451 (1964).

26. *See* cases cited in 56 *Am. Jur. Waters* §§ 476, 477, pp. 891–895; III *American Law of Property*
(Casner ed., 1952) § 15.26, pp. 855–856; *Black's Law Dictionary* (rev. 4th ed., 1968) 36–37, 102,
173–174, 529, 530, 637, 1455, 1594; *Ballentine's Law Dictionary* (3d ed., 1969) 14, 64, 116, 339, 414,
1085, 1229; 6 Powell, *The Law of Real Property* (3d ed., 1939) (hereinafter cited as "Powell") ¶ 983,
pp. 607–611; 5A Thompson, *Commentaries on the Modern Law of Real Property* (1957 Repl.) (hereinafter cited as "Thompson") §§ 2560–2563, pp. 599–619; 4 Tiffany, *The Law of Real Property* (3d ed.,
1939) (hereinafter cited as "Tiffany") § 1219, pp. 613–615.

27. *See, generally,* 2 Shalowitz, *Shore and Sea Boundaries* 536–539 (1964).

and Geodetic Survey as evidence in this case. We think the maps should be given full credence, and should be taken as absolutely establishing the truth of all that they purport to show . . . . Capt. George R. Campbell, United States engineer and hydrographic surveyor, testified to the accuracy of official hydrographic maps, stating that all the features connecting the shores with the water are accurately outlined and surveyed and tied to permanent landmarks, that these surveys are made with extreme accuracy, and that all are worked on an astronomical basis and are chained and taped a number of times, and that the government is always careful to do as accurate work as is possible on a coast line and in its marine coast survey work. Such testimony was hardly necessary, we think, for the court might properly take judicial notice of the accuracy of the official plats of the United States Coast and Geodetic Survey.

The California Supreme Court has written to the same effect: "The Court is invited to and will take judicial notice of the coast lines within the state and will also resort to publications issued by the Department of Commerce describing and delineating the United States coast and geodetic surveys."[28]

While maps depicting the surveys of the U.S. Coast & Geodetic Survey may be judicially noticed, an adversary may be permitted to introduce expert testimony tending to show inaccuracies of the survey. But the latitude allowed such experts is circumscribed:

Defendants asked certain questions of the engineers who were witnesses on their behalf, which called for the opinions of these gentlemen concerning the accuracy and reliability of the Coast & Geodetic Survey maps heretofore referred to, for the purpose of defining boundaries in private ownership which by deed or statute is defined to be the low-tide line. Objections to these questions were sustained, and properly so. These questions called for opinions of the witnesses as to the weight to be given by the jury to these maps as evidence. If any inaccuracies existed in the maps themselves these experts, if qualified to do so, could have been permitted to point them out; and they were permitted to do exactly this; but they could not be permitted to give their general opinions as to the value of these maps for the purpose of defining boundaries, nor as to the weight to be given them as evidence. These were questions for the jury, whose province it was to determine the weight and value of the maps as evidence. These same experts were permitted to spend much time pointing out the

---

28.   *Boone v. Kingsbury,* 206 Cal. 148, 273 P. 797, 813 (1928). *See also Los Angeles v. Duncan,* 130 Cal.App. 11 (1933). Decisions from other states are to the same effect. *See, e.g., Van Dusen Inv. Co. v. Western Fishing,* 124 P. 677 (Ore. 1912); *Rockaway Pacific Corp. v. State,* 203 N.Y.Supp. 279, 286 (1924), *aff'd per curiam, Rockaway Pacific Corp. v. State,* 154 N.E. 603 (1926).

incompleteness of these maps, and this is as far as they should have
been permitted to go.[29]

Judicial notice has also been taken in certain cases that a river or lake is navigable.[30]
Title to the beds of navigable lakes and rivers in general passed to the states upon their admission to the Union.[31]

Other documents of relevance to land and boundary cases that have traditionally been the
subject of judicial notice include topographic maps of the Department of Interior,[32] as well as
reports of the California Debris Commission and of the United States Bureau of Reclamation [33]

Deeds are to be construed by the laws in force at the time they were executed; that is a
well-settled proposition.[34] A 1953 California decision held, "[a]ny question as to the title to
property, acquired prior to the conquest, is to be determined by reference to the Mexican law
at the time the property was acquired. Any question, however, as to restraint or restrictions
upon the use of property, acquired prior to the conquest, is to be determined according to
American law."[35] For these purposes, courts have taken judicial notice of Spanish and Mexican laws in force prior to the annexation of California by the United States.[36]

An appellate court may take judicial notice of all matters properly subject to judicial notice, such as land patents, government reports, and so on, even when the trial court had declined to do so.[37]

Judicial notice has also been taken of the records of a variety of state agencies.[38] Certain official acts of municipal entities are judicially noticeable as well. California Evidence
Code section 452(b), for example, provides that judicial notice may be taken of "regulations
and legislative enactments issued by or under the authority of . . . any public entity in the
United States." Under this principle, the courts of some states have judicially noticed a
number of maps prepared by city or county officials.[39] Some courts, however, particularly in

---

29. *City of Oakland v. Wheeler,* 34 Cal.App. 442, 452–453 (1917).

30. *See People v. Gold Run D & M Co.,* 66 Cal. 138 (1884); *Los Angeles v. Aitken,* 10 Cal.App.2d 460
(1935); *Donnelly v. United States,* 228 U.S. 708 (1913).

31. *Shively v. Bowlby,* 152 U.S. 1 (1894).

32. *Newport v. Temescal Water Co.,* 149 Cal. 531 (1906); *Union Transp. Co. v. Sacramento County,* 42
Cal.2d 235, 239 (1954).

33. *Gray v. Reclamation Dist. No. 1500,* 174 Cal. 622 (1917).

34. *Elder v. Delcour,* 269 S.W.2d 17, 23 (Mo. 1954), and *Dell v. Lincoln,* 102 N.W.2d 62, 68 (Neb. 1960).

35. *Hart v. Gould,* 119 Cal.App.2d 231, 236 (1953).

36. *City of Los Angeles v. City of San Fernando,* 14 Cal.3d 199 (1975); *Venice Peninsula Properties v.
City of Los Angeles,* 31 Cal.3d 288 (1982), *rev'd on other grounds, Summa Corp. v. California,* 466 U.S.
198 (1984).

37. *White v. State of California,* 21 Cal.App.3d 738, 762 (1971); *Chas. L. Harney, Inc. v. State of California,* 217 Cal.App.2d 77, 85–86 (1963).

38. *See, e.g.,* cases collected in 31 Cal.Jur.3d, *Evidence* § 36, at 72–73 (1976).

39. *Carrollton R.R. v. Municipality,* 19 La. 62 (1941) (city map admitted to evidence); *Adams v. Stanyan,*
24 N.H. 405 (1852) (town map made under state authority admitted); *Blackman v. Riley,* 138 N.Y. 318, 34
N.E. 214 (1893) (ancient map made by a city surveyor for a private party received); *but see Dobson v.
Whisenhant,* 101 N.C. 645, 8 S.E. 126 (1888) (unofficial map inadmissible); *Burwell v. Sneed,* 104 N.C.
118, 10 S.E. 152 (1889) (map made by a surveyor appointed by the county excluded); *Cowles v. Lovin,* 135
N.C. 488, 47 S.E. 610 (1904) (certificates of survey by a former county surveyor then in Texas excluded).

California, have refused to take judicial notice even of municipal or city ordinances or the rules of local boards.[40]

So we have considered the kinds of documents of which the courts have taken judicial notice. Bearing in mind the rationale of the hearsay rule, it is now appropriate to consider whether the truth of matters asserted in such documents are judicially noticed equally with the documents themselves. That is to say, it is one thing to take judicial notice of the charts and descriptive reports of the United States Coast Survey; it is another to take judicial notice of the *truth* of all matters asserted within those documents. In *Beckley v. Reclamation Board,* the California Court of Appeal in 1962 wrote:

> It is, of course, true that the courts take judicial notice of all matters of science and common knowledge, and of the reports of the California Debris Commission referred to above; that is to say, we take judicial notice of the fact that the reports were made, and of their contents. We do not, however, take judicial notice that everything said therein is true. These reports are based upon studies made by engineers with opinions and conclusions drawn from those studies. But engineers are not infallible, nor are all statements contained in the reports, even those stated as facts, irrefutable. (If they were, then there could have been no justification for the Grant Report which modified the Jackson Report, corrected errors therein, and changed the physical plan [for a flood control project] in several material respects where experience and further studies had proved earlier engineers' assertions of "fact" and their conclusions faulty. . . .) To assert the immutableness of statements in official documents would constitute abdication by the courts in favor of adjudication by engineering fiat. [Citations omitted.][41]

(But now isn't that wholly at odds with the Ninth Circuit's 1919 decision in *United States v. Romaine,* discussed a few pages back?)

Needless to say, the distinction between taking judicial notice of an official document, and of taking notice of the truth of matters stated in the documents, is frequently lost. Courts often make findings of fact based solely on items contained in judicially noticed official documents. Thus, as mentioned at the beginning of this chapter, courts have found that the moon arose at a particular time because it was so reported in an official document.

When the distinction is rigidly maintained, a serious problem may arise if the only source for a critical fact in a case is a statement, a statistic, or an observation in a government report or study. The report may be judicially noticed as an official act, but its various statements are hearsay and may not be admissible for the truth of the matter asserted. (The reader may ask whether there are appropriate exceptions to the hearsay rule in a given situation.) The truth of a statement in a government report such as "The water-surface elevation of Lake

---

40.   *See South Shore Land Co. v. Petersen,* 226 Cal.App.2d 725, 746 (1964), and cases cited.
41.   205 Cal.App.2d 734, 741–742. For the same proposition, *see People v. Long,* 7 Cal.App.3d 586, 86 Cal.Rptr. 590 (1970); *Marocco v. Ford Motor Co.,* 7 Cal.App.3d 84, 86 Cal.Rptr. 526 (1970); B. Witkin, *California Evidence* § 180, at 167 (2d ed., 1966).

Tahoe reached 6,230.5 feet in 1944" perhaps may not be judicially noticed, however. But the truth of the statement may find its way into the record of the trial through other principles of evidence. For example, an expert witness might base an opinion in part on the statement, providing that the necessary predicates are shown.

Curiously, it has been noted that "nowhere can there be found a definition of what constitutes competent or authoritative sources for purposes of verifying judicially-noticed facts."[42]

Courts as a rule do not take judicial notice of private maps and instruments of conveyance between private parties, even when they are of record.[43] But while judicial notice may not be taken under such circumstances, there are other rules of evidence that may achieve the same result. California Evidence Code section 1600, for example, provides the following:

> (a)   The record of an instrument or other document purporting to establish or affect an interest in property is prima facie evidence of the existence and content of the original recorded document and its execution and delivery by each person by whom it purports to have been executed if:
> (1)   The record is in fact a record of an office of a public entity [such as the County Recorder's Office]; and
> (2)   A statute authorized such a document to be recorded in that office.

Even when a court takes judicial notice of the contents of an official survey, however, it confines itself to recognizing only those matters that lay within the authority of the government surveyor. Justice Story wrote for the United States Supreme Court in an early decision:

> The survey, made by a surveyor, being under oath [of office] is evidence as to all things which are properly within the line of his duty. But his duty is confined to describing and marking on the plat the lines, corners, trees, and other objects on the ground, and to subjoin such remarks as may explain them; but in all other respects, and as to all other facts, he stands, like any other witness, to be examined on oath in the presence of the parties, and subject to cross examination. . . . But it has never been supposed that if in such a survey the surveyor should go on to state collateral facts, or declarations of the parties, or other matters, not within the scope of the proper official functions, he could thereby make them evidence as between third persons.[44]

---

42.   Comment, The Presently Expanding Concept of Judicial Notice, 13 Vill. L. Rev. 528, 545 (1968).
43.   *See South Shore Land Co. v. Petersen,* 226 Cal App.2d 725, 746 (1964). *See also Wooster v. Butler,* 13 Conn. 309 (1839) (survey must be one made by authority of law); *Doe v. Roe,* 35 Del. 229, 162 A. 515 (1930), in which the issue was the location of marshland. Government maps and a pamphlet describing the soil conditions were excluded in a decision characterized by Professor Whitmore as "unsound; no authority cited." *Simmons v. Spratt,* 20 Fla. 495, 499 (1884) ("the simple filing of a private survey in a public office does not make it evidence").
44.   *Ellicott v. Pearl,* 35 U.S. (10 Pet.) 412, 441 (1836).

Numerous other matters beyond the scope of discussion permissible in this book have also been the subject of judicial notice, such as various historical and geographical facts, the identity of principal officers of the federal and state governments, and such mundane matters as the identity of the court's own officers.[45]

A phenomenon that has been much criticized but that is nonetheless recurring with some regularity is the massive use of judicial notice in the form of the so-called Brandeis brief. This expression refers to the practice of lawyers incorporating in their appellate briefs a vast amount of socioeconomic or political data from ostensibly empirical studies, none of which was subject to verification at the trial-court level through the process of cross-examination. A well-known example of this wholesale use of the principle of judicial notice is found in the landmark decision of *Brown v. Board of Education*.[46] In *Brown*, the United States Supreme Court held that racially segregated schools cannot meet the requirements of equal protection of the law, notwithstanding their equality of teachers or facilities, because the very act of segregation brands the segregated minority with a feeling of inferiority. The briefs of Brown and the decision of the court all made rife use of published socioeconomic studies of the effect of segregated schools to support the holding that segregated schools are inherently "unequal." Virtually all of this evidence was received through the vehicle of judicial notice. Without a protracted discussion of the phenomenon of the Brandeis brief, it may do to observe that its principal criticism is that such data rarely meet the primary criterion for the application of judicial notice—that the information to be judicially noticed is relatively "indisputable."[47]

A final matter to be considered briefly is whether, after judicial notice has been taken of a fact, a party may introduce evidence contradicting that fact. Theoretically, this problem arises from the notion that to be judicially noticeable, a fact must be "indisputable." That being the case, any evidence tending to contradict it would be inherently incredible and not entitled to be admitted. This point of view has been expounded by some of the ablest legal writers and accepted by a number of courts.[48] A contrary view holds that the principal function of judicial notice is to expedite the trial of cases, and that judges should dispense with the need for time-consuming formal evidence when the fact in question is most likely (as opposed to indisputably) true. When this argument is accepted, then it follows that evidence contradicting the judicially noticed fact may be admitted and that the finder of fact may then accept or reject the truth of the fact which had been judicially noticed.[49]

---

45. *See, e.g.,* cases collected in McCormick on Evidence § 330, at 765, (2d ed., 1972).

46. 347 U.S. 483 (1954).

47. For further reading on this subject, see Davis, "Judicial Notice," 55 Col. L. Rev. 945 (1955); Note, "Social and Economic Facts-Appraisal of Suggested Techniques for Presenting Them to the Courts," 61 Harv. L. Rev. 692 (1948); and for some enlightened thoughts on the proper role of the courts in fashioning social policy, see Cardozo, *The Nature of the Judicial Process,* 113–125 (1921); and Frank, *Law and the Modern Mind,* Ch. 4 (1930).

48. See the authorities collected in McCormick on Evidence § 332, at 769, n. 4 (2d ed., 1972).

49. See authorities collected in McCormick on Evidence § 332, at 769, n. 5 (2d ed., 1972).

Chapter 9

The Opinion Rule and Expert Testimony

"The public buys its opinions as it buys its meat, or takes in its milk, on the principle that it is cheaper to do this than to keep a cow."

—Samuel Butler, *The Notebooks* (London: Jonathan Cape, 1926), p. 263

Ordinarily a witness must be shown to have personal knowledge of the matter he is to testify about (this personal knowledge is sometimes called the "foundational fact"), before he testifies about it. And he must confine his testimony to what he has actually perceived. Inferences from those perceptions are for the jury to draw. Stated conversely, the general rule is that a witness may not testify about his opinions. This rule is an ancient one; at common law, witnesses were to testify only to "what they see and hear."[1] Wigmore/quotes Lord Coke in 1622: "[i]t is no satisfaction for a witness to say that he 'thinketh' or 'persuadeth himself'."[2]

Legal commentators have written a great deal on what constitutes an "opinion" (by which is meant a conclusion or inference), and of course on whether there ought to be a rule excluding such evidence. Is it an opinion, for example, to say that a man was drunk? Is it an opinion to say that a fence was not fit to keep stock off of land?[3] The distinction between "fact" and "opinion" is a rarified one and, on reflection, appears perhaps to be purely artificial. To say there were "remnants" of a fence post is to draw a conclusion from certain observations; the same applies if one were to say that the fence post was of redwood; or that it was in a state of decomposition from water saturation and not having been painted with preservative;

---

1. Phipson, *Evidence,* 398 (9th ed., 1952), quoting from Anon. Lib. Assn. 110, 11 (1349). *See also* 9 Holdsworth, History of English Law 211 (1926).
2. 7 Wigmore on Evidence, § 1917 at 2 (3rd ed., 1940) quoting from *Adams v. Canon,* Dyer 53b.
3. Such testimony was held inadmissible in *Baltimore & O.R. Co. v. Schultz,* 43 Ohio 270, 1 N.E. 324 (1885).

or, that the monument set in 1922 was to mark a property corner. In formulating his conclusion in each of these examples, the speaker considers not only what he actually saw, but also additional data from his personal store of knowledge. (First, he recalls and considers how fence-post remnants appear; second, he considers what redwood looks like; third, the effects of not treating buried wood; and fourth, the acts of a certain land surveyor in 1922.) With the possible exception of what partly rotted wood looks and feels like, much of this knowledge may not be shared by the population at large. (That is to say, were another to see—not "look at" but "see"—what the witness saw, the testimony of that person might well be different.)

In the last statement of the hypothetical witness given above, he is uttering what most people "intuitively" know to be "opinion," as opposed to fact. But to truly relate mere fact, in the sense of only what was observed, the speaker would have to convey raw sensory data—perhaps not even in words—because there are no words that can convey only what was perceived, unencumbered by data learned elsewhere (e.g., that an object with certain visual and palpable characteristics is wood). Until we learn directly to transmit raw sensory data from the observer to the trier of fact, it is perhaps helpful to bear in mind that the distinction between fact and opinion is an artificial, though useful, one.

The reader who has had even a minimum of exposure to courtroom proceedings may recall numerous items of testimony that fall into this opinion-fact netherland. A 1937 Alabama decision devoted an extensive discussion to the admissibility of a witness's testimony that "he looked like he was dying."[4] A Texas court similarly grappled with the testimony that "the situation at the end of Pearl Street presented such an appearance that a stranger on a rainy night would be liable to drive off into the river."[5] In a 1903 Wyoming decision, it was held proper to allow a witness to testify whether the defendant's admission of killing the victim was a "sincere" or "joshing" remark.[6] In the common case of a disputed location of a property corner, where there are no eyewitnesses to its setting, it is safe to assume that any testimony as to the location of the corner will be considered opinion testimony.

The place of opinion testimony in boundary cases has received a good deal of treatment by the courts and commentators over the years. The exhortations of Judge Cooley to the land surveyor, made in his now-famous nineteenth-century address, might be considered here, before turning to consideration of the specific principles that govern the giving of expert testimony. In that address, "The Judicial Functions of Surveyors," Cooley said:

> Of course nothing in what has been said can require a surveyor to conceal his own judgment or to report the facts one way when he believes them to be another. He has no right to mislead, and he may rightfully express his opinion that an original monument was at one place, when at the same time he is satisfied that acquiescence has fixed the rights of parties as if it were at another. But he would do mischief if he were to attempt to "establish" monuments which he knew would tend to disturb settled rights; the farthest he has a right to go as an officer of the law is to express his opinion where the

---

4. *Pollard v. Rogers,* 234 Ala. 92, 173 So. 881 (1937).
5. *City of Beaumont v. Kane,* 33 S.W.2d 234, 241, 242 (Tex.Civ.App. 1930).
6. *Horn v. State,* 12 Wyo. 80, 148, 73 P. 705, 721–723 (1903).

monument should be at the same time that he imparts the information to those who employ him, and who might otherwise be misled, that the same authority that makes him an officer, and entrusts him to make surveys, also allows parties to settle their own boundary lines, and considers acquiescence in a particular line or monument for any considerable period as strong, if not conclusive, evidence of such settlement. The peace of the community absolutely requires this rule. *Foyce v. Williams,* 26 Mich. Reports, 332. It is not long since that in one of the leading cities of the State an attempt was made to move houses two or three rods into a street, on the ground that a survey, under which the street had been located for many years, had been found on a more recent survey to be erroneous.[7]

The one traditional exception to the rule forbidding opinion testimony was the testimony of an expert's opinion, when certain conditions were met. (The modern rules of expert testimony are discussed below.) As a concession, however, to the impossibility of distinguishing absolutely between fact and opinion, many jurisdictions have seized the dilemma's horns and provide that nonexpert opinion testimony may sometimes be given. For example, Federal Rule of Evidence 701 provides the following:

> If the witness is not testifying as an expert, his testimony in the form of opinions or inferences is limited to those opinions or inferences which are (a) rationally based on the perception of the witness and (b) helpful to a clear understanding of his testimony or a determination of a fact in issue.

A number of states have nearly identical provisions with respect to nonexpert opinion testimony.[8] The reasons for the modern rule expressed in Rule 701 are well-expressed in these remarks of the Advisory Committee on the Proposed Federal Rules of Evidence, which urged its adoption for the federal courts:

> The rule retains the traditional objective of putting the trier of fact in possession of an accurate reproduction of the event.
> Limitation (a) is the familiar requirement of first-hand knowledge or observation.
> Limitation (b) is phrased in terms of requiring testimony to be helpful in resolving issues. Witnesses often find difficulty in expressing themselves in language which is not that of an opinion or conclusion. While the courts have made concessions in certain recurring situations, necessity as a standard for permitting opinions and conclu-

---

7.   Cooley, "The Judicial Functions of Surveyors," reprinted in Carhart, *A Treatise on Plane Surveying,* pp. 403–404 (1888).

8.   *See, for example,* Cal. Evid. Code § 800; Kansas Code of Civ. Proc. § 60-456(a); N.J. Evid. Rule 56 (1); and Uniform Rule of Evid. 56-1.

sions has proved too elusive and too unadaptable to particular situations for purposes of satisfactory judicial administration. Moreover, [there is] the practical impossibility of determining by rule what is a "fact,". . . . The [proposed] rule assumes that the natural characteristics of the adversary system will generally lead to an acceptable result, since the detailed account carries more conviction than the broad assertion [i.e., an opinion or conclusory statement of the witness], and a lawyer can be expected to display his witness to the best advantage. If he fails to do so, cross-examination and argument will point up the weakness. If, despite these considerations, attempts are made to introduce meaningless assertions which amount to little more than choosing up sides, exclusion for lack of helpfulness is called for by the rule. [Citations omitted.]

A.   EXPERT TESTIMONY

The traditional exception to the common law's prohibition on opinion evidence, as has been noted, was the testimony of experts. Such testimony was originally permitted only in the rarest of circumstances, when the expert's opinion was grudgingly felt essential to a decision. Today the rules respecting the admissibility of an expert's testimony, whether in the form of an opinion or otherwise, are codified in most jurisdictions. It is convenient to consider two codifications of these principles. The Federal Rules of Evidence, for example, provide the following:

> If scientific, technical, or other specialized knowledge will assist the trier of fact to understand the evidence or to determine a fact in issue, a witness qualified as an expert by knowledge, skill, experience, training, or education, may testify thereto in the form of an opinion or otherwise.[9]

California Evidence Code section 801, which is similar to the rules of most states, provides the following:

> If a witness is testifying as an expert, his testimony in the form of an opinion is limited to such an opinion as is:
> (a) Related to a subject that is sufficiently beyond common experience that the opinion of an expert would assist the trier of fact. . . .

AN EXPERT'S NON-OPINION TESTIMONY
One noteworthy distinction between the Federal and California rules is that the federal rule provides that the expert "may testify . . . in the form of an opinion *or otherwise.*"

---

9.   Fed. R. Evid. 702.

Certainly a land surveyor, when he has been qualified as an expert, and when the other predicates have been shown, may testify that in his opinion the section corner in question was originally set at a certain location. It is certainly proper for him on other occasions to testify to the customary methods of land surveyors in seeking to recover original corners, the contents of applicable instructions from the General Land Office to the United States deputy surveyors, and so on. Notwithstanding the nearly metaphysical problems associated with the distinction between opinion and fact, such matters are not what the law considers "opinions." Yet they may be of even more value in enabling the trier of fact to reach a decision than the ultimate opinion of an expert such as a land surveyor. In many cases it may be unnecessary for the land surveyor even to render an opinion; it may suffice for him simply to relate to the court some segment of his store of specialized knowledge. The Advisory Committee on the Proposed Federal Rules of Evidence explained it this way:

> Most of the literature assumes that experts testify only in the form of opinions. The assumption is logically unfounded. The rule accordingly recognizes that an expert on the stand may give a dissertation or exposition of scientific or other principles relevant to the case, leaving the trier of fact to apply them to the facts. Since much of the criticism of expert testimony has centered upon the hypothetical question, it seems wise to recognize that opinions are not indispensable and to encourage the use of expert testimony in non-opinion form when counsel believes the trier can itself draw the requisite inference. The use of opinions is not abolished by the rule, however. It will continue to be permissible for the experts to take the further step of suggesting the inference which should be drawn from applying the specialized knowledge to the facts.[10]

The California rule, though silent on the point, would not seem to preclude the use of such non-opinion testimony by expert witnesses. Moreover, there appear to be no California decisions purporting to preclude such testimony. Actually, it would seem that the use of such non-opinion testimony by experts should be welcomed by the courts, because it would obviate the often-voiced fear that juries may be unduly impressed with the opinion on an ultimate question in the case given by a highly qualified expert in the field.

A PRE-CONDITION: THE NEED FOR THE EXPERT TESTIMONY

Whether an expert will be permitted to testify depends in the first instance on the need of the trier of fact to hear such testimony. "There is no more certain test for determining when experts may be used than the common sense inquiry whether the untrained layman would be qualified to determine intelligently and to the best possible degree the particular issue without enlightenment from those having a specialized understanding of the subject involved in the dispute."[11] Both Federal Rule 702 and Section 801 of the California

---

10. Advisory Committee Comments to Proposed Rule 702.
11. Ladd, *Expert Testimony,* 5 Vand. L. Rev. 414, 418 (1952).

Evidence Code contain this requirement. When expert testimony, whether in the form of opinion or not, is excluded it is usually because it is unhelpful, and therefore a waste of the court's time.[12]

THE EXPERT'S QUALIFICATIONS: A "FOUNDATIONAL FACT"

To testify as an expert, whether to matters of opinion or merely to matters within his specialized knowledge, such as scientific procedures, a witness must first be shown to be "qualified" to give such testimony. In a typical formulation of the required qualifications of an expert, Federal Rule 702 specifies that the requisite expertise may be the product of "knowledge, skill, experience, training or education." Most states describe expert qualifications in similar terms.[13] Under these criteria, it is certainly not a difficult matter to qualify a witness as an expert.[14]

The qualifications of an expert are normally elicited on direct examination by the attorney who called the witness, but it is necessary to do so only if the opposing party objects to the offered expert testimony; in the absence of an objection the testimony may be received without a showing of the witness's qualifications. As a matter of tactics, however, it is usually imprudent to introduce expert testimony, even if one can get away with it, without a showing of the witness's qualifications, because the trier of fact will know nothing of the witness by which it can ascribe credibility and weight to his testimony.

As a matter of procedure, the qualification of an expert is a "preliminary" or "foundational" fact that must be determined by the judge, who is vested with a broad discretion in making such determinations.[15] The party offering the witness as an expert has the burden of proving that the expert is qualified (that is, unless the opposing party has acquiesced to the witness's testimony by failing to object.) Ordinarily the jury will not be informed of the fact that the court has determined the witness to be qualified as an expert.[16] Obviously, a jury is not bound to accept the opinion of an expert, and the weight to be given to the opinion is solely within its province.[17]

In determining whether a witness is qualified to testify as an expert, the trial judge typically considers the following kinds of factors: (1) the number of experts in the field, (2) whether the type of testimony involves generally recognized and related scientific fields, (3) whether claims of accuracy by the witness have been substantiated by accepted methods of scientific verification, (4) whether the expert's background, whether based on education, practical experience, and so on, is sufficient to enable him to testify, and (5) whether the expert's background relates sufficiently to the type of testimony to be given.[18]

---

12. *See* 7 Wigmore on Evidence § 1918.
13. *See, e.g.,* Cal. Evid. Code § 720.
14. The relative ease with which such qualifications can be established is shown by the California cases of *Moore v. Belt,* 34 Cal.2d 525 (1949); *Hyman v. Gordon,* 35 Cal.App.3d 769, (1973); and *Brown v. Colm,* 11 Cal.3d 639 (1974).
15. Cal. Evid. Code § 405; see *Putensen v. Clay Adams, Inc.,* 12 Cal.App.3d 1062, 1080 (1970); *People v. Murray,* 247 Cal.App.2d 730, 735 (1967).
16. Cal. Evid. Code § 405(b).
17. *See, e.g.,* California Jury Instructions, Civil, 5th ed., 36 No. 2.40.
18. *See People v. King,* 266 Cal.App.2d 437, 443–445 (1968), in which voice prints and testimony relating to them were rejected.

It has been held, for example, that a psychologist may not ordinarily testify about his interpretations of an electroencephalogram or about the psychological effects of drugs because of his lack of medical training,[19] and a chemist may not testify that something is of reasonable medical certainty.[20] Land and surveying experts may expect their testimony to be similarly restricted. Proffered testimony of a land surveyor as to the customary practices of lending institutions, for example, will most likely be found to be beyond his experience and, thus, inadmissible.

An expert need not be academically trained in the subject matter of his expertise to be qualified to testify; the expertise may be the product of skill, experience, or training obtained outside of an academic environment. In one California case, the testimony of an FBI expert who did not have a university degree, but who did have considerable experience in fingerprint identification, was allowed.[21] In another, a product salesman was permitted to testify as to the value of the product, even though he was not an appraiser by profession.[22] In addition, there is no requirement that one be licensed to engage in the business in which he is expert in order to give expert testimony. It is conceivable, for example, that one who for whatever reason has never troubled himself to obtain a state license to practice land surveying may be an expert in corner recovery.

THE BASES FOR AN EXPERT'S OPINION

Another topic of expert testimony that has received much attention is what matters an expert may properly use as the basis for his opinion testimony. The California statute provides that such testimony may be:

> Based on matter (including his special knowledge, skill, experience, training, and education) perceived by or personally known to the witness or made known to him at or before the hearing, whether or not admissible, that is of a type that reasonably may be relied upon by an expert in forming an opinion upon the subject to which his testimony relates, unless an expert is precluded by law from using such matter as a basis for his opinion.[23]

Federal Rule 703 is somewhat simpler:

> The facts or data in the particular case upon which an expert bases an opinion or inference may be those perceived by or made known to him at or before the hearing. If of a type reasonably relied upon by experts in the particular field in forming opinions or inferences upon the subject, the facts or data need not be admissible in evidence.[24]

---

19. *See People v. Davis,* 62 Cal. 2d 791, 801 (1965).
20. *See* Holmes, *Trials of Dr. Coppolino,* pp. 228–230.
21. *People v. Stuller,* 10 Cal.App.3d 582, 597 (1970).
22. *Naples Restaurant, Inc. v. Coberly Ford,* 259 Cal.App.2d 881, 884 (1968).
23. Cal. Evid. Code § 801(b).
24. Fed. R. Evid. 703.

Essentially, then, the expert may base his opinion on three sources of information. First, he may consider what he has perceived with his own senses. A treating physician testifying at trial is an example of an expert relying on what he has himself perceived. Likewise, a land surveyor who has made a field survey has personal observations on which he may base his testimony.

The second basis consists of facts made known to an expert at or before the hearing. This kind of information is frequently presented to the expert in the form of the much-maligned "hypothetical question." In this situation, the expert witness is asked to assume as true a series of facts (evidence of which must, as a general rule, have been received in court) and then to render his opinion on the basis of those assumptions.

Finally, the expert may rely on matter that "reasonably may be relied upon by an expert" in his area. A physician, for example, may rely on statements made to him by his patient concerning the history of his condition.[25] In appropriate cases, he may also rely upon the reports and opinions of other physicians.[26] An expert on real-property values may rely on inquiries he has made of other people, commercial reports, market quotations, and relevant sales of which he has learned, for the purpose of rendering an opinion on value of property being condemned.[27]

This last category is perhaps the most significant one. It should be noted that, under this category, an expert may base his opinion testimony on matters that, under other circumstances, would not be admitted in evidence because they constitute hearsay in the preceding example. For example, all of the information obtained by the expert on real property values from other persons, commercial reports, and so on, is hearsay. Are there any restrictions on an expert's use of hearsay as a basis of his testimony? The only general rule that governs this issue is the test whether it is acceptable practice in the expert's field to rely on such statements.[28]

On what kinds of matter may a land surveyor reasonably rely in his practice of land surveying and thus use as a basis for his testimony in court? There are certain indispensable matters, such as plats, field notes, contracts, and special instructions for original government surveys, which are considered reasonable bases of opinions. Recognized texts (such as Bouchard & Moffitt) likely fall into this category as well. In most circumstances, however, the determination must be made on a case-by-case basis. If the surveyor is attempting to locate a bearing tree noted in the field notes of an 1869 survey, and he finds nothing but new-growth trees in the terrain, may he rely on community "reputation" of a devastating forest fire in 1940? May he rely on newspaper accounts of such a fire to account for his inability to find the bearing tree? Similarly, in an accretion–avulsion case, may he rely upon newspaper or diary accounts of a great flood to form a basis for his opinion as to the manner in which a

25.   *People v. Wilson,* 25 Cal.2d 341 (1944).
26.   *Kelley v. Bailey,* 189 Cal.App.2d 728 (1961).
27.   *Betts.v. Southern Cal. Fruit Exchange,* 144 Cal. 402 (1904); *Hammond Lumber Co. v. County of Los Angeles,* 104 Cal.App. 235 (1930); *Glantz v. Freedman,* 100 Cal.App. 611 (1929).
28.   These determinations must be made on a case-by-case basis. *Board of Trustees v. Porini,* 263 Cal.App.2d 784, 793 (1968).
29.   The author has from time to time suggested that land surveyors compile a manual of guidelines for such purposes, which should be based on the historical and customary practices of their profession. Such a manual, when it receives general recognition among land surveyors and perhaps even by state agencies that license land surveyors, would go far toward answering some of these questions.

river changed its course?[29] These are the kinds of questions that need to be pondered in advance by both the attorney and the expert witness preparing to testify. It is largely a question of custom among experts in the particular field, and it is impossible to generalize about the kinds of matter that will be permissible bases for opinions in any particular case.

The same principle holds with respect to whether an expert may base his opinion in part on the opinion of another expert. Thus, whether a land surveyor may base his opinion in part upon the expert opinion of a dendrochronologist on the age of certain trees is governed by the criterion whether a land surveyor ordinarily may rely upon such an opinion.[30]

AN EXPERT'S OPINION BASED ON SCIENTIFIC TECHNIQUE—MUST THE TECHNIQUE HAVE "GENERAL ACCEPTANCE IN THE FIELD"?

The federal courts—and the nearly 40 States that have adopted the Federal Rules of Evidence—had for years held to the view that expert opinion based on scientific technique was inadmissible unless the technique was "generally accepted" as reliable in the relevant scientific community.[31] The United States Supreme Court overturned that rule in *Daubert v. Merrell Dow,* 509 U.S. 579 (1993), after several federal appeals courts had begun to reject the so-called *Frye* rule.

The *Frye* rule derived from a case concerning the admissibility of evidence from a systolic blood-pressure deception test. In 1923, this was a crude precursor to the polygraph machine. The *Frye* Court of Appeals wrote:

> Just when a scientific principle or discovery crosses the line between the experimental and demonstrable stages is difficult to define. Somewhere in this twilight zone the evidential force of the principle must be recognized, and while courts will go a long way in admitting expert testimony deduced from well-recognized scientific principle or discovery, the thing from which the deduction is made must be sufficiently established to have gained general acceptance in the particular field in which it belongs.[32]

The Supreme Court in *Daubert* held that the Federal Rules of Evidence, adopted by the Supreme Court in 1972 and effective July 1, 1973, overturned the *Frye* rule. The court's words are worth setting out verbatim:

> Here there is a specific Rule [Federal Rule of Evidence] that speaks to the contested issue. Rule 702, governing expert testimony, provides:
> "If scientific, technical, or other specialized knowledge will assist the trier of fact to understand the evidence or to determine a fact

---

30. *See, e.g., People v. Wilson,* 25 Cal.2d 341, 348 (1944). Compare *San Bernardino County Flood Control Dist. v. Sweet,* 255 Cal.App.2d 889, 902 (1967).

31. *See, e.g., Frye v. United States,* 54 App.D.C. 46, 47, 293 F. 1013, 1014 (D.C. Cir. 1923).

32. 293 F. at 1014.

in issue, a witness qualified as an expert by knowledge, skill, experience, training, or education, may testify thereto in the form of an opinion or otherwise."

Nothing in the text of this Rule establishes "general acceptance" as an absolute prerequisite to admissibility. Nor does Respondent present any clear indication that Rule 702 or the Rules as a whole were intended to incorporate a "general acceptance" standard.

The Supreme Court was quick to point out, however, that its ruling did not mean that every crackpot's view must be admissible as "opinion" evidence:

> That the Frye test was displaced by the Rules of Evidence does not mean, however, that the Rules themselves place no limits on the admissibility of purportedly scientific evidence. Nor is the trial judge disabled from screening such evidence. To the contrary, under the Rules the trial judge must ensure that any and all scientific testimony or evidence admitted is not only relevant, but reliable.[33]

THE EXAMINATION OF THE EXPERT WITNESS, AND THE HYPOTHETICAL QUESTION

The direct examination of an expert witness—that is, the questioning of him by the lawyer who called him to the stand—usually proceeds along the following lines:

1. "Mr. Webster, will you please describe your background for the court?" (The witness's testimony concerning his qualifications, as with the rest of his testimony, should be well-rehearsed with his attorney. If the list of honors, achievements, and so on, of the witness is particularly long, it is better that the attorney ask him a series of questions relating to his academic background, his work experience, membership in professional societies, publications, and so on, rather than have the witness hold forth for a lengthy period of time, and perhaps appear to be unduly immodest.)

2. "Have you been retained by me to investigate the matter that is the subject of this lawsuit?" (Typically, there would follow a series of questions inquiring when the witness was first contacted by the lawyer, what he was asked to do, etc.)

3. "After you learned of the issues presented by this matter in the meeting at my office, what if anything did you do to investigate?" (Recall that without "if anything" the question would assume a fact not yet in evidence—that is, that the expert did perform some investigation. The expert would then describe any field surveys, library research, computations, and so on, that he has made as part of his initial investigation.)

4. "As a result of your investigations which you have described, have you come to any conclusions concerning the question you were asked to investigate? (The answer should be a simple "Yes". The statement of the expert's conclusion should await the following question.)

---

33.   509 U.S. 588–589.

5.   "And what is your conclusion [or opinion]?" (At this point, the expert should state in the briefest possible terms his conclusion.)

6.   "Would you please explain the reasons and basis for your opinion in this respect?" (It is at this point that the expert may draw a deep breath, and explain why he concluded as he did. This process may take 30 minutes, or several days, of court time. If this explanation is lengthy, it is helpful to have the attorney direct it by means of questions interspersed at regular intervals. This technique helps break up a discourse that may be growing tedious, and it helps to summarize one point of the explanation and lead into the next, better enabling the jury to follow the testimony.)

The hypothetical question, mentioned above, is a traditional, though much criticized, method of eliciting an expert's opinion on the witness stand.[34] This is the method of asking the witness to assume a series of propositions as true, and then asking his opinion. Hypothetical questions are a proper, though frequently undesirable, method of presenting evidence.[35] The assumptions that the expert witness is asked to accept must each be founded on some evidence that has already been received. If the hypothetical question asks in part, for example, "And assume further that a 100-year flood occurred in this watershed in 1939," there must already have been some evidence of such a flood; otherwise the question is objectionable. A hypothetical question need not include as assumptions all evidence that has been received, and it may omit unfavorable evidence; if the unfavorable evidence is later believed by the jury, of course, the opinion testimony given in answer to the hypothetical question is to that extent impaired. Also, the hypothetical question must state as assumptions all facts essential to the expert's opinion.[36]

When an attorney wishes to elicit an expert opinion by means of a hypothetical question, it is common for him to write out the question in advance and study it with the expert prior to his testimony. In this way they can be assured that no essential facts have been omitted, and the expert can aid the attorney in formulating the question to make it more accurate and to make his answer more convincing.

On the cross-examination of an expert who has rendered his opinion in response to a hypothetical question, it is a familiar technique to ask the expert, without the benefit of the written hypothetical question before him, to restate all of the facts that he deems essential to his opinion. Then, one by one, the expert is asked whether his opinion would be different if a particular fact were other than as the expert was asked to assume. In this and in a number of other respects, the cross-examination of an expert is much broader than that of an ordinary witness. Typically, an expert witness may be cross-examined as to his qualifications, the general subject to which his expert testimony relates, the matter upon which his opinion is based (whether testified to on direct examination or not), and the reasons for his opinion.[37] In addition, the compensation and expenses paid to the expert by the party calling him to the stand

34.   *See* B. Witkin, *California Evidence* §§ 1179–1181 (2d ed., 1966)
35.   Generally, it is preferable that an expert has arrived at his conclusion through personal observation of the critical facts.
36.   *See,.e.g., Waller v. Southern Pacific Co.,* 66 Cal.2d 201, 210 (1967).
37.   *See, e.g.,* Cal. Evid. Code § 721; Fed. R. Evid. 705 *People v. Nye,* 71 Cal.2d 356, 374 (1969), *cert. den.,* 406 U.S. 972 (1972).

is a proper subject of cross-examination.[38] Many attorneys believe that this line of investigation is ordinarily not productive, while others feel that the disclosure of the high fees of an expert may serve to implant a suspicion of the witness's motives in the minds of the jurors.

THE CROSS-EXAMINATION OF AN EXPERT WITNESS

A lawyer may cross-examine an opposing expert witness in the same manner as any other witness may be cross-examined. In addition, the lawyer may cross-examine the witness about the general subject of his or her testimony and, of course, the expert's qualifications. California law used to be that an expert could not be cross-examined with regard to the "content or tenor" of any technical book or publication in the expert's ostensible field, unless the expert had relied on the book or publication in forming the opinions at issue. In 1997, California came into line with the majority of the other states, as well as the federal system, on this question. Now, like the Federal Rules of Evidence (which are the rules in at least 39 states as well as the federal system), California law provides that an expert may be cross-examined with respect to a text or publication in the expert's field if the publication is established as a reliable authority by the testimony or admission of the witness or by other expert testimony or by judicial notice."[39]

MISCELLANEOUS OBSERVATIONS ON EXPERT TESTIMONY

Contrary to an ancient and discredited yet frequently repeated dictum, it is permissible for an expert to give his opinion on an "ultimate issue" in the case. Thus if the question in a trial is the true location of the original corner between sections 1, 2, 11, and 12, a qualified land surveyor may offer his opinion as to its location, notwithstanding that the issue is the ultimate question in the case.[40]

Questions asking, "Is it possible that . . .," while once of dubious propriety when posed to an expert, are now accepted in most jurisdictions as proper.[41]

Many court decisions have addressed the proper scope of expert testimony to be given in land cases by surveyors and civil engineers.[42] In a California case requiring the location of the "high-water mark" boundary, the expert testimony of land surveyors that the water of the river had washed a smooth, visible line along the base of the river bank, below which it was devoid of vegetation, was held competent evidence.[43] On the other hand, it has been held that a trial court erred in admitting the testimony of a surveyor that an arch projected less than 3 inches over the boundary line of a lot, when the survey was not actually made by the witness but rather by his employees, and when he was present only for a few minutes at the beginning of the survey, which took six or seven hours.[44] It is not possible to generalize about the circumstances in which an expert may testify on the basis of work performed by subordi-

---

38. *See* Cal. Evid. Code § 722.

39. Cal. Evid. Code § 721 (b); Fed. R. Evid. 803 (18).

40. *See* Fed. R. Evid. 704 (a); Cal. Evid. Code § 805.

41. *See, e.g., Bauman v. San Francisco,* 42 Cal.App.2d 144 (1940); *In Re J.F.,* 268 Cal.App.2d 761 (1969); *Cullum v. Seifer,* 1 Cal.App.3d 20, 26 (1969).

42. *See, e.g., Richfield Oil Corp. v. Crawford,* 39 Cal.2d 729 (1952).

43. *Mammoth Gold Dredging Co. v. Forbes,* 39 Cal.App.2d 739 (1940).

44. *Hermance v. Blackburn,* 206 Cal. 653 (1929).

nates, but this situation is always potentially troublesome and should be considered in advance of trial. Broadly speaking, the practice of having such work done by subordinates and the degree of supervision are salient factors in determining the admissibility of the testimony.

In general, the testimony of a competent surveyor who, in surveying the location of a line, followed the rules of the federal land office is admissible to show the location of the disputed boundary line of a government section.[45] Also, as may seem obvious, the testimony of a surveyor concerning the location of lands within a section is not made inadmissible simply because he did not survey the entire section, when no such survey was essential to locate the lands in question.[46]

There are, as in all fields of law, anomalous cases. Once court remarked, erroneously it seems clear, that opinion evidence of the location of a boundary line is inadmissible. (That court did, however, hold that allowing such testimony was not sufficient reason for reversal.[47]) Another court has similarly held that it was "harmless error" to allow the admission of testimony of a surveyor that his survey was correct and another was not.[48] (For an explanation of the concept of "harmless error," see Chapter 13, regarding the appeal of cases.)

Surveyors, it should be added, are not the only witnesses able to testify about the location of boundary lines. Thus, the location of a township line need not be established solely by the testimony of surveyors.[49] Laymen who assisted in the making of such a survey may testify about the location of stakes set[50] and the location of landmarks placed by the surveyors.[51]

This chapter would be incomplete without at least a few remarks of practical advice to the land expert preparing to testify in court. It is a common (and sound) practice for an attorney cross-examining an expert witness to ask very narrow questions—usually requiring a yes or no answer. A question that is so broad as to ask "why" the witness holds such an opinion, what the basis is, and so on, provides the witness with an excellent opportunity to expand on his answers and hammer home or clarify points he realizes may have been made imperfectly earlier. The persuasive effect of such testimony is amplified when it comes out in response to questions by an adverse attorney. Such opportunities are rare and should be seized on.

When the witness is compelled to answer yes or no, or to give a similarly restricted response to a narrow question, he should rarely, if ever, request permission of the judge to expand on his answer. He will only appear to be too much of an advocate for his client's position; in any event, his attorney can ask for elaboration on re-direct examination.

When the witness does not know the answer to a question, he should admit so without hesitation.

When he has been asked a "yes or no" question, and it is apparent that the question is not susceptible of a yes or no answer, the reason is almost invariably that the question assumes a fact that either is not true or is not in evidence. If the witness's attorney does not perceive the predicament and object to the question (as he ought to, unless as a tactical matter

45. *Porter v. Counts*, 16 Cal.App. 241 (1911).
46. *Heinlen v. Heilbron*, 97 Cal. 101 (1892).
47. *Andrews v. Wheeler* 10 Cal. App. 614, 618 (1909).
48. *Tognazzini v. Morganti*, 84 Cal. 159 (1890).
49. *Heinlen v. Heilbron*, 97 Cal. 101 (1892).
50. *Anderson v. Richardson*, 92 Cal. 623 (1892).
51. *Kimball. v. McKee*, 149 Cal. 435 1906.

he would rather have the witness fend for himself and not appear to require his protection), it is often an effective rejoinder to inform the cross-examiner (and the court) that his question reminds you of the question, "Have you stopped beating your wife?" By explaining the infirmities in the question, the witness may also reiterate or reaffirm the points he wants to make.

It may be true, as some have asserted, that our litigation system has become largely "trial by expert witness." The late San Francisco trial attorney Melvin Belli wrote, "[I]n the courtroom, science now aids man more than ever before in his search for truth. In modern trials, while the lay witness testifies to what he saw, or, more correctly put, that he believes he saw, the expert witness, with cold, calculable facts, figures, and photographs, answers objectively to the question 'What really happened?' "[52] Even Belli, however, professed remarkable optimism that these battles of the experts produce no less just results than before:

> When the personal injury case gets to court, there will be plaintiff's medical expert to meet defendant's medical expert. While defendants will testify that plaintiff is ready to "go twenty rounds with Rocky Marciano," plaintiff's doctor will gravely shake his head and opine that plaintiff's in such horrible shape he should be in a bottle on the shelf at Harvard Medical School!
>
> These facetious illustrations may be quite all right to contemplate abstractly, you say, but "if I'm injured I don't want my own lawyer to exaggerate my claims, neither do I want defendant insurance company grossly to minimize them. I want the truth!"
>
> Well, the jury will get that truth (justice) for you. That defendant's doctor gazing benignly at the jury and saying, in his soothing, syrupy voice "There's nothing wrong," isn't taking in that jury as much as you may fear. Nor will your doctor, if he exaggerates, be able to sell a bill of goods to them either.
>
> Amazingly enough, that jury, with absolutely no medical, no forensic, no legal experience, but with an awful lot of collective common sense, 99 times out of 100 will give a proper evaluation.[53]

---

52. Belli, "Ready for the Plaintiff!" 207 (1956).
53. *Id.* at 206.

Part II

The Procedure of a Civil Case

Chapter 10

The Pleading and Motion Stages

*The attitude of early English chancery courts [or courts of equity] towards pleadings of inexcusable prolixity may be seen in the case of Richard Mylward. In that case, the filing of a pleading amounting to six score sheets of paper which "might have been well contrived in sixteen sheets of paper," so outraged the court that, in addition to imposing a fine upon the pleader, it ordered that the Warden of the Fleet take the pleader into custody and "bring him into Westminster Hall, on Saturday next, about ten of the clock in the forenoon, and then and there shall cut a hole in the myddest of the same engrossed replication * * * and put the said Richard's head through the same hole * * * and then, the same so hanging, shall lead the same Richard, bare headed and bare faced, round about Westminster Hall, whilst the Courts are sitting, and shall shew him at the bar of every of the three Courts within the Hall * * *."*

—Punishment of Richard Mylward for drawing, devising, and engrossing
a replication of the length of six score sheets of paper (1596), Monro,
Acta Cancellariae (1817) 692, noted in 5 Holdsworth,
History of English Law (1924) 233.

A lawsuit is begun by filing with the court a document called the "complaint."[1] The person filing the complaint—the one beginning the lawsuit or "action"—is usually called the "plaintiff," sometimes "petitioner"; the person being sued is the "defendant" or "respondent." The complaint must state clearly what relief the plaintiff seeks (e.g., money damages or a court decree compelling the defendant to do or refrain from doing some act), as well as facts that, if true, would legally entitle the plaintiff to the relief he seeks. The allegations, of course, may or may not be true. Determining the truth of the allegations is the function of the trial, not the pleadings.

---

1. *See, e.g.,* Cal. Civ. Proc. Code § 411.10.

What are "pleadings"? They are the initial court papers that frame the issues in a case. They include a plaintiff's complaint, the defendant's "answer," the defendant's cross-complaint or counterclaim, and answers to such countersuits.[2]

A lawsuit to "quiet title" to real property provides a simple illustration of what things a complaint must state or "plead." In a quiet-title action the plaintiff seeks a court decree confirming his title. The complaint need only plead three things: (1) that plaintiff is the owner of the disputed property (which must be adequately described), (2) that the defendant or defendants claim some interest in the subject property adverse to the interest of the plaintiff, and (3) that the interest claimed by the defendant is without any basis,—i.e that is, that defendants in fact have no right, title, or interest in the disputed property adverse to plaintiff's interest. The "prayer" for relief in a quiet-title action asks that the court enter a decree "quieting" plaintiff's title against the claims of defendants and prohibiting or "enjoining" the defendants from asserting their claims in the future.

Quiet title is a "form of action," that is, a type of lawsuit.[3] There are other forms of action in which the title to or boundaries of real property may be placed in issue.

A complaint in "ejectment," for example, is somewhat similar to a complaint in quiet title, except that the plaintiff in ejectment is asking for more than a declaration of title; he is also seeking to regain possession of the disputed property. The necessary allegations for a complaint in ejectment are similar to those for a complaint in quiet title except that the plaintiff must also allege that the defendant or defendants are in "wrongful" possession of the subject property. By "wrongful" possession is meant one "adverse" to the title of the plaintiff. A lessee who is complying with the terms of his lease is, in this sense, not "possessing adversely" to his lessor, even though the lessor may consider his tenant's occupation undesirable or "adverse" in a nonlegal sense, because the tenant is obnoxious or because the lease provides for below-market rents.

The concept of "adverse possession," familiar to many as "squatters' rights," originally developed and is still employed as a defense to an ejectment action.[4] An ejectment action will fail if the plaintiff does not bring it within the applicable "period of limitations," which varies from state to state from as few as 5 years to as many as 20.[5] If the defendant can show that he has been in "wrongful" possession of the property for a period greater than the limitations period, he will not be ejected from the premises and may in fact have title quieted in him, by virtue of his adverse possessions.

To sue for trespass to real property, which is another form of action, the plaintiff must allege that he owns the disputed property and that the defendant wrongfully (i.e., without permission or other legal right) entered it, for example, by merely walking on it. If this is all that

---

2. *See, e.g.,* Cal. Civ. Proc. Code §§ 420, 422.10.

3. Federal law now provides for quiet-title actions against the United States; these were once banned under the doctrine of sovereign immunity. *See* 28 U.S.C. § 2409a; *Block v. North Dakota,* 461 U.S. 273 (1983). California's once-scattered and confusing welter of laws relating to quite-title actions are now codified in Code of Civil Procedure §§ 760.010 through 771.020.

4. The matters that are to be pleaded as defenses are discussed below in the paragraphs concerning "answers."

5. It is 5 years in California, Code of Civil Procedure sections 315–320. Actions to quiet title against the United States must be brought within 12 years of the date plaintiff knew or should have known of his claim. 28 U.S.C. 2409a(g).

is alleged, plaintiff is entitled to prevail, but he will be awarded only "nominal" damages, say one dollar. Naturally, few cases are brought for such trifling matters, but the usual case involves something more substantial—some timber has been cut, gravel or minerals have been removed, and so on. Thus, in trespass cases the defendant will frequently assert as a defense that the plaintiff doesn't own the disputed property or that the common boundary between their properties is such that the activity complained of actually occurred on defendant's land.

A form of action that was unknown in common law but that has been adopted in many jurisdictions and become popular as a vehicle for resolving various real property disputes is the action for "declaratory relief."[6] Using this form of action, the plaintiff simply alleges that a dispute has arisen between the defendant and him, and he requests the court to spell out in a decree the rights and obligations of the parties with respect to the dispute. A declaratory relief action may be used to determine many kinds of rights and duties with respect to real property, as well with respect to numerous other subjects, such as a contract, a previous court decree, and so on. Moreover, the pleading requirements are simple. The plaintiff merely alleges that an "actual controversy" has arisen between the plaintiff and the defendant and sets forth the particulars of the controversy. For example, the plaintiff may allege that a controversy has arisen concerning the location of the common boundary line between property owned by himself and property owned by the defendant. He then sets forth his own position as to the correct location of the boundary line and the position he understands defendant to take. The prayer in a complaint for declaratory relief asks the court to resolve the dispute by a decree specifying the rights of the parties. In a boundary dispute, for example, the decree would specify the common boundary between the parties.

The same principles of pleading, it might be added, apply to lawsuits outside the real-property field. In a lawsuit to recover money damages for personal injuries suffered in an automobile accident, for example, the plaintiff must plead specific facts to recover under one or more legal theories. If he is suing the automobile manufacturer, he would employ different theories than if he were suing the driver of the vehicle that struck him, and naturally the requirements vary according to the legal theory used.

There are a number of legal theories that impose liability for personal injuries. The most common, "negligence," is based on fault. To sue on a theory of negligence, the plaintiff must plead specific matters that, if true, would render the defendant liable. Other theories have been developed that impose liability without requiring that defendant be shown to have been negligent. Some of these theories have developed from the recognition that certain kinds of activities inherently carry a great risk of harm (the operation of an explosives factory, for example), and thus as a matter of policy one who undertakes such an activity ought to be held "strictly liable" for injuries caused by his activity without the plaintiff having to show negligence or carelessness. Another such theory is premised on the notion that to prove negligence in a given case is virtually impossible, yet the injury clearly could not have occurred without negligence. The classic case is of a flour barrel falling from a warehouse window onto the plaintiff; the theory is called *"res ipsa loquitur"*[7]—"the thing speaks for itself."

---

6. *See, e.g.,* Cal. Civ. Proc. Code §§ 1060–1062.5.
7. *Byrne v. Boadle,* 2 H. & C. 722, 159 Eng. Rep. 299 (1863). "If that phrase had not been in Latin, nobody would have called it a principle." Lord Shaw, in *Ballard v. North British R.C.* (1923), Sess. Cas., H.L., 43.

Other theories of "no fault" or absolute liability are frequently applied in cases of defective products, such as drugs. Each of these theories has specific elements that must be pleaded in the complaint, just as the various forms of action for real-property disputes require the pleading of specific elements.

To explore the matter of terminology one further step, take the case of an injury to real property, by explosion for example. The injured party is said to have one "cause," or right of action—that is, the right to bring and pursue a lawsuit. Thus, a plaintiff whose factory was destroyed by an explosion in an adjacent factory has one "cause of action" for that injury, although he may have several theories, or "counts," on which to base a monetary recovery. Trespass is one possible theory, because the courts might well view the energy and debris emanating from the defendant's factory as being tantamount to the defendant's having intentionally taken an axe and sledgehammer to plaintiff's factory. Other possible theories are liability for conducting an ultrahazardous activity or for simple negligence. Were the plaintiff himself injured in the explosion, he would have another, separate cause of action for his personal injuries; again he may have several theories of recovery, or "counts," to plead in his complaint for this cause of action. (Not all lawyers and judges, it should be noted, adhere to this distinction—sometimes referred to as the "Pomeroy theory" of primary rights—between a "cause of action" and a "count"; it is a useful one, however.)

Let us focus on lawsuits in which the title to or boundaries of real property are in controversy. When the complaint has been filed, it is customary for the plaintiff to record in the recorder's office of the county in which the property is located a "notice of lis pendens," which means, "notice of pending action." The purpose of this notice, which is sometimes called simply a "lis pendens," is to notify any interested persons, and particularly any purchaser of the property, that the lawsuit is pending. Anyone purchasing the property will then take title "subject to" the lawsuit; in other words, while he has not personally been sued, he will be bound by the outcome in the case nevertheless. To protect the title he believes he has purchased, the purchaser should substitute himself in the lawsuit for his seller once the sale has closed. The lis pendens also protects the plaintiff from such a purchaser, later arguing that he had no notice of the lawsuit and claiming that he is, therefore, not bound by the judgment in the case. Customarily a certified copy of the lis pendens is filed with the county or federal court clerk, who keeps the official court records of all papers filed in the lawsuit.

At the time the plaintiff files his complaint, he obtains from the county clerk a "summons," which together with the complaint is then "served" upon the defendant. The summons is the court's exercise of its power to bring the defendant to the court, where the defendant will thus be bound by the decision in the case. This power is often called the court's "*in personam*" jurisdiction—that is, its jurisdiction over the person of the defendant.[8]

This "service" of the complaint and summons, then, renders the defendant subject to the court's jurisdiction. Until the defendant has been served with a copy of the summons and complaint, or voluntarily appears in the action, proceedings in the lawsuit, including a final judgment, cannot affect him.

---

8. A court's jurisdiction can also be viewed from the standpoint of whether the court has jurisdiction over the type of case—for example, whether it is empowered to hear quiet title cases, or jurisdiction over the geographic area. Can a California court ever hear a case of disputed title to land in Nevada, for example?

Service may be accomplished in a variety of ways. The complaint and summons may be personally served on a defendant, frequently by the sheriff, simply by handing them to the defendant. In most jurisdictions today, the complaint and summons can be served by mail. Defendants who cannot be found may be served by "publication"—for example, by publishing a notice of the lawsuit in a newspaper. The problems, including constitutional ones, associated with the use of this form of service are too numerous to mention.

A defendant who has been served with a copy of the summons and complaint typically has 30 days or some comparable period in which to respond. Several courses of action are open to him. He may, for example, take the position that service of the complaint and summons upon him was somehow improper or ineffective and move, therefore, to "quash" the service. If this motion is successful, it is as though he had never been served, and future proceedings in the lawsuit can have no effect on him until he is properly served.

Another alternative for the defendant is to "demur." At common law and in most states, a demurrer is a procedure the defendant may employ if the complaint is ill-pleaded. If a complaint in quiet title, for example, neglects to plead that defendant's claim of title is without any right, this defect may be raised by demurrer.[9] If the court "sustains" the demurrer, the complaint is stricken, although the plaintiff is ordinarily given a period of time in which to file an amended complaint that is properly pleaded. If it appears to the court that under no circumstances could the plaintiff amend his pleading to properly state a cause of action, the court will sustain the demurrer "without leave to amend." This act effectively ends the lawsuit at the trial level, leaving the plaintiff with appeal as his only recourse. Under the Federal Rules of Civil Procedure, the objection that the plaintiff has not pleaded facts sufficient to constitute a cause of action—called a "claim for relief" in the federal system—is made by a "motion to dismiss." Like the demurrer, this procedure in the federal system may also be used to call the court's attention to several other kinds of defects in the complaint.[10]

While there are several other steps that the defendant may take when he has been served, typically he will file his "answer" to the plaintiff's complaint.[11] This document, when read together with the plaintiff's complaint, "joins the issues." In the answer, the defendant addresses himself to the allegations made by the plaintiff in his complaint. He does this by admitting or denying those allegations or by making new allegations of his own. In a quiet-title action, for example, the defendant will no doubt admit that he asserts a claim to the subject property adverse to the claim of plaintiff. On the other hand, he will deny that the plaintiff is the owner of the subject property and similarly deny that his own claims are without foundation. Frequently a defendant feels that it is inadequate to simply admit or deny a particular allegation, and in these instances he responds to plaintiff's allegation with an allegation of his own. If a plaintiff, for example, has pleaded that he holds a valid deed from defendant to the subject property, the defendant may simply deny the allegation or may allege something affirmatively, such as the fact that the deed was delivered under duress or was forged.

The answer also affords the opportunity for the defendant to plead "affirmative defenses," which are typically defined as "any new matter tending to defeat the claim of

---

9. *See, e.g.,* Cal. Civ. Proc. Code § 430.10.
10. Fed. R. Civ. P. 12.
11. *See, e.g.,* Cal. Civ. Proc. Code §§ 430.10–431.70.

plaintiff." In a real-property lawsuit, the defendant may wish to allege that the action brought by the plaintiff is barred by a statute of limitations[12] or by the doctrines of laches or estoppel. These matters would appropriately be pleaded as affirmative defenses. Laches is a defense that typically applies in so-called "equitable" (as opposed to "legal") actions where no statute of limitations is specifically applicable. It alleges, in general, that the plaintiff has unduly delayed in bringing his case. The doctrine of equitable estoppel holds that when one party has made representations to another (such as that the other is the owner of certain lands), and the second party has relied on these representations, the first party will not be permitted to deny the truth of what he had formerly asserted. It is a doctrine often sought to be employed in land-title cases.[13]

Motions for "summary judgment" or "summary adjudication," or for partial summary judgment or adjudication, have become exceedingly popular in the past 25 years, although no one can explain why. These motions require nearly as much or more effort to prepare than the time required to actually try the case. And judges are singularly reluctant (some unkind souls have said "afraid"—afraid of reversal on appeal) to grant these motions.

A motion for summary judgment says, basically, that the case presents no genuine issue of material fact.[14] If the plaintiff is bringing such a motion in a case alleging adverse possession, he must adduce written statements of witnesses attesting to the basic elements of the cause of action—possession for the requisite number of years, payment of taxes, and so forth. The theory is that if the defendant cannot refute these affidavits with affidavits of his own, so that the material fact becomes "genuinely disputed," judgment should be entered in favor of the plaintiff. Occasionally, whole cases are disposed of on summary judgment.

Obviously, when there is any real dispute as to a material fact, summary judgment—as to the whole case or any part of it—is inappropriate. The procedure is finding a useful renascence, though, in modern environmental litigation, when the lawsuit is predicated on the defendant's own publicly filed reports. In these cases, such as under the Federal Clean Water Act, the defendant's reports are admissions. Summary judgment is thus an expedient tool to (a) establish violations of the particular statute and (b) move the case on to the penalty phase.[15]

---

12. A quiet-title suit against the United States, for example, must be brought within 12 years of the date the plaintiff knew or should have known of the claim of the United States. 28 U.S.C. § 2409a (f); *See California v. Arizona*, 440 U.S. 59 (1979).
13. *See e.g., City of Long Beach v. Mansell*, 3 Cal. 3d 462 (1970).
14. Fed. R. Civ. P. 56; Cal. Civ. Proc. Code § 437c.
15. *See, e.g., Sierra Club v. Union Oil Co. of Cal.*, 813 F.2d 1480, 1491–1492 (9th Cir. 1988), *vacated for reconsideration*, 485 U.S. 931 (1988), *reinstated and amended*, 853 F.2d 667 (9th Cir. 1988).

Chapter 11

The Age of Discovery

"Question: And you are involved, I take it, in both aspects of pathology here in your practice?"

"Answer: Yes."

"Question: Directing your attention to the 6th of November, 1976, in the evening hours, do you recall performing an autopsy on a man by the name of Robert Edgarton at the funeral chapel?"

"Answer: Yes."

"Question: And Mr. Edgarton was dead at that time, is that correct?"

"Answer: No, you dumb a—h—. He was sitting on the table wondering why I was doing an autopsy on him."

> —From transcript of deposition (with deceased's name changed and expletive
> partially deleted) taken by a member of the San Francisco Bar,
> sometime in the 1970s.

"Discovery" is a generic term used to describe a number of pretrial procedures that enable each party to learn about his opponent's case. Through discovery, a party can learn (a) the contentions his opponent makes, (b) facts and witnesses known to the opponent, and (c) documents in the opponent's possession. The discovery phase is particularly important to land experts, because it is frequently during this phase that they are consulted by one of the parties and become one of the participants in the discovery process.

Formerly, civil lawsuits culminated in what came to be known as the odious "trial by surprise." Parties to a lawsuit endeavored to keep secret whatever they knew of their case. They could learn of the opponent's case only by stealth, such as by hiring private investigators. At trial, one hoped he had anticipated enough of his opponent's case—what evidence he would introduce, what witnesses he would call—and conversely that he could surprise his opponent with a witness or document for which the opponent had no ready answer. The concern arose that this system tended to decide cases not so much on the merits as on the degree of prestidigitation displayed by the lawyers. As a consequence, in the 1940s and 1950s many jurisdictions, including the federal courts, began to allow for broader pretrial discovery procedures. The new policy was to see that a litigant went to trial fully prepared not only to present his own case, but also to meet the case of his opponent. Ideally, each would know what witnesses the opponent would call, what they would testify about, and what documents would be introduced. This state of affairs was designed to make trials shorter and fairer, with the results based more on the merits than on the wizardry of the lawyers. Also, it was anticipated that these more liberal discovery laws would result in a greater number of cases settling out of court: When each side has a thorough knowledge of the strengths and weaknesses of the other's case, it is more likely that they will be able to compromise.

As might be expected, while the element of surprise has been reduced to a substantial degree, a trial still in no way resembles a well-rehearsed play in which each actor knows his own lines, the lines of the other actors, and the precise sequence of scenes. Evidence or testimony a party relies upon to establish his case may be ruled inadmissible by the judge at trial. A witness may recall events differently from the way he recounted them during the discovery phase of the case. New documents or witnesses may be found on the eve of trial, after discovery has been concluded.[1] Nevertheless, it is fair to say that the advent of broad pretrial discovery in civil actions has resulted in more cases settled, fairer trials, and, perhaps, shorter trials. (While the time of the average civil trial appears to have elongated in the past 20 years or so, the discovery rules are likely not at fault. By inducing parties to agree to certain facts that are virtually indisputable—the authenticity of a township plat, the dates of land sales, and so on—many hours and even days of trial time are saved.)

What follows is a brief discussion of the most commonly used discovery devices. Others, such as physical examinations of a plaintiff in a personal injury action, are not of importance to the expert in natural-resource cases. It should be emphasized that to be effective, discovery must be conducted meticulously. Several examples will help to show how the expert working with his attorney can help ensure that discovery yields truly helpful information and is not squandered.

A.   INTERROGATORIES

Interrogatories are written questions one party asks of (or "propounds to") another party in the lawsuit, which must be answered under oath.[2] (They may not be sent to a witness, for example, who is not a party.) They may inquire whether the party makes a certain contention: "Do you contend that the corner common to Township 3 South, Range 2 West, and Township 4 South, Range 3 West, Mt. Diablo Base and Meridian, as set by the Deputy United States Surveyor, has been obliterated?" Interrogatories may also ask what facts the opponent relies upon for making a particular contention: "If your answer to the preceding in-

---

1. Typically, discovery must be concluded by a specified time prior to trial, such as 30 days. An attorney may purposely delay the search for witnesses and documents until after discovery has been concluded, so that the opponent has no opportunity to discover these matters. This problem is readily avoided by a pretrial order, for example, that all case preparation be concluded 60 days, and discovery 30 days, prior to trial. Or the order might provide that any new evidence or witnesses found after discovery has concluded must be disclosed to the other party.

2. *See, e.g.,* Cal. Civ. Proc. Code § 2030; Fed. R. Civ. P. 33.

terrogatory is in the affirmative, please describe all facts that you contend support your contention that the corner has been obliterated." Interrogatories may also inquire whether the opponent knows of any documents pertinent to the lawsuit. "Please identify all documents that you contend support your contention that the corner has been obliterated."

Interrogatories are also used to learn the names and addresses of witnesses known to the other party: "Please identify by name and last known address all persons with knowledge of the of the negotiations leading to the sale of Blackacre by Mr. Green to Mr. Snowden."

In land cases, the expert is frequently asked to help draft interrogatories to the opposing party and to respond to interrogatories served ("propounded" again is the usual expression) by the opposing party. A cooperative effort by both the lawyer and the expert generally will produce the best results; the lawyer will be most familiar with the legal issues and principles, while the land expert will know more of the technical, scientific, and practical aspects of the particular subject matter. Both are needed in preparing useful interrogatories. Take the interrogatory concerning the township corner used as an example above, and suppose it read, "Do you contend that the corner common to Sections 1, 2, 11, and 12 [the description should be precise so that there is no question what corner is referred to] is lost or obliterated?" Your opponent responds, "Yes." At this point you do not know whether the opponent contends the corner is obliterated, or whether he contends it is lost. This problem may be the result of poor drafting (usually the fault of the attorney) or of ignorance of the distinction between a lost and an obliterated corner (about which the expert could have educated the attorney).[3] In other instances, the land expert may be able to suggest whole lines of inquiry that, otherwise, would remain unknown to the lawyer.

Of interrogatories, and of all discovery mechanisms in general, it can be said that the scope of permitted inquiry is far broader than the limits of relevance at trial. This concept may be termed "discovery relevance" for a shorthand expression, and it is discussed in somewhat further detail later in this chapter. While it is no objection to a discovery inquiry (such as an interrogatory) that the question calls for hearsay, information that is protected by a privilege is protected from disclosure during discovery to the same extent as during trial.

B.   DEPOSITIONS

A deposition is a discovery device by which the oral testimony of a witness, under oath, is taken before a shorthand reporter, who later transcribes the testimony. A deposition may be taken of a party or of a nonparty witness (so long as the witness is within the geographic limits of the court's subpoena power). It is usually taken in the office of the attorney who wishes to take the deposition. Unlike trial, there is no judge present to rule on objections during a deposition.

A deposition, like interrogatories, also differs from courtroom testimony in that the scope of relevance is much broader than at trial. The witness may even give hearsay testimony that would be inadmissible at trial, so long as the testimony is "relevant to the subject matter of the action" or "calculated to lead to the discovery of admissible evidence."[4]

---

3. *See, generally, State of California v. Thompson,* 22 Cal. App.3d 368 (1971).
4. *See, e.g.,* Cal. Civ. Proc. Code §§2017(a), 2030(f); Fed. R. Civ. P.30.

Occasionally an impasse develops during a deposition; the witness's attorney instructs the witness not to answer a particular question, and the attorney asking it insists on an answer. Because no judge is present, such disputes must be taken to court for a decision whether the witness must respond. Occasionally, the deposition is interrupted at this point to allow the attorneys to argue the matter before a judge promptly, sometimes in a matter of days. If the question is not critical to the balance of testimony, however, the deposition may continue on other points, with the dispute to be argued later, or the disputed question may be ignored. The witness, of course, should always follow his attorney's instruction when told not to answer.

It is not often that an expert being deposed will be instructed by his attorney not to answer a question. Such an instruction not to answer is usually based on the assertion that the information asked for is privileged, and ordinarily the only such privilege attaching to information in the expert's possession is the "attorney's work product privilege." But for most purposes that privilege is waived when the expert is identified as a potential witness for trial, and his discussions and correspondence with his attorney are usually then subject to discovery. It is more often the case that a witness is instructed not to answer when the witness is the client himself, and the question concerns information protected by the attorney–client privilege. On rare occasions, a witness will be instructed not to answer because the inquiry has strayed beyond the bounds of discovery relevance.

Invariably, surveyors and other experts who are to testify in a land case will be deposed before trial. (Obviously it can be rather hazardous for an expert to testify at trial if the corresponding expert who will testify for the opposing party has not been deposed. The best expert in any field can never be certain he has overlooked nothing, that there is no plausible way other than his own of interpreting the evidence.) The deposition of an expert is taken— that is, the questions are asked—by the opposing attorney. The expert's own attorney will seldom ask him questions during a deposition, because he has nothing to "discover."

The deposition of any expert witness generally covers a predictable range of topics, although not necessarily in predictable order. The witness is asked his qualifications—his education, membership in professional associations, his number of years in practice, work experience, professional journals to which he subscribes, and so forth. If the case concerns the location of a water boundary, the deponent may be asked how many such problems he has encountered during his career. If the witness is a civil engineer, and the case concerns property boundaries, he is frequently asked what percentage of his practice is devoted to land surveying.

Next the witness will typically be asked about his substantive preparation for trial: (1) Specifically, what has he been asked to do in connection with the case? (2) In carrying out his assignment, what materials has he reviewed, and what investigations has he made? (3) What are his conclusions (opinions)? (4) What reasons does he have for his conclusions?

Obviously, in a complicated case, a thorough deposition of a land expert, which fully covers these four broad areas, may take days. The response to the simple question, "What materials have you reviewed in the course of your work on this case?" may alone take many hours. Almost invariably the witness is required to bring to his deposition all such materials (or copies of them). The attorney deposing the witness, if he is thorough, will ask the witness to identify each document, map, spreadsheet, and so on, and to explain its contents and relevance to the case. Each item will be given an exhibit number (Plaintiff's Exhibit 1, etc.) and made part of the record of the deposition.

Preparing to testify at a deposition is as essential as preparing to testify at trial. The expert will want to adduce every fact and document he has examined and be able to articulate clearly the reasons for his opinions. If the attorney takes the deposition skillfully, he will take care to press the witness at every point for a definitive answer: "Are there *any* other materials you have reviewed in the course of your investigation which you have not identified? . . . Are you certain? . . . If you later recall any you may have forgotten today, will you notify me promptly?" Or, "You have stated three bases for your conclusion. Do you have any other reason for your conclusion? Be certain now—these three only?" If at trial the surveyor produces a survey he did not have at his deposition or states an additional reason for his conclusion, he may be in for some rough moments. The opposing attorney may try to make it appear the new information was deliberately withheld or recently fabricated.

On the other hand, the deposition may be conducted in an unskillful or unthorough manner. In this regard, the witness should recall that he is obligated only to answer the questions asked. If the attorney neglects to inquire into a critical point or to press for the witness's reasoning, the witness has no obligation to volunteer the information. For this reason most attorneys instruct their own witnesses who are about to be deposed to listen carefully to the question asked and answer *that question* only.

Many additional practical tips may be offered to the expert preparing for his deposition. Such advice often should be adapted to suit the circumstances of the case as well as the styles of both the expert and his attorney. A typical letter of advice to the witness about to be deposed, setting forth many such tips, is contained in Appendix 1.

The expert is frequently—and prudently—asked to help his attorney prepare to take the deposition of the opposing party's expert. The attorney may have little or no experience with technical issues presented by the case and, thus, may need a healthy dose of education. On the other hand, he may be well-versed in the mathematics entailed, but have little knowledge of the historical practices of the General Land Office, for example, or the rules respecting the restoration of lost corners, the valuation of nonuse resources, or the method of capitalizing income. Even when the attorney is knowledgeable about the subject, he will usually find the suggestions of his expert helpful, much as he would the suggestions of his law partner on the procedural or tactical aspects of the case. The expert, of course, should always suggest that the attorney inquire further into those areas that the expert finds troubling or unclear. The deposition will probably be the only occasion before trial to learn what the opponent's expert has found or concluded on the subject.

Depositions, of course, are primarily a discovery device—that is, a technique to learn what a witness, whether a party or not, may know about the case or about matters affecting the case. They have other uses too, however. Ordinarily a deposition may not be read at trial, since the witness himself should testify. But if a witness contradicts his deposition testimony, he may be confronted with it. Too, if the witness has died or is otherwise unable to be produced for trial (as when he lives beyond the court's subpoena power and refuses to come voluntarily), on proof of his unavailability his deposition may be used at trial.

C.   REQUESTS FOR ADMISSIONS

The request for admissions is another discovery tool the expert may be asked to help prepare. It is frequently the last-used discovery device and is designed to eliminate uncontroverted

ˏ issues of fact and expedite the conduct of trial. Like a set of interrogatories, a request for admissions is a written document sent to the opposing party. It may request that the opponent admit the genuineness of a document (recall that a document must be shown to be authentic before it may be introduced into evidence), or the truth of a statement of fact. More modernly, it may be used to request the admission of an opinion, or of the applications of law to facts.[5]

As is explained in Chapter 2, "Documentary Evidence," the first requirement for the admissibility of a document is a showing that it is genuine—for example, that it in fact is the plat of the survey in question, the letter it purports to be, and so on. Even in a relatively uncomplicated boundary case, there may be scores of documents that a land surveyor and his attorney will seek to introduce at trial. The request for admissions may obviate the need for proof of the genuineness of these documents at trial. Typically, the attorney assembles copies of the documents he wishes to introduce at trial and labels them "Exhibit A," "Exhibit B," and so forth. The request for admissions, which precedes the package of documents, then reads something like this:

> *Request for Admission No. 1.* Do you admit that the document attached hereto as Exhibit A is a true and correct copy of a letter from Simon Simpson to the Commissioner of the General Land Office dated January 27, 1897?
>
> *Request for Admission No. 2.* Do you admit that the document attached hereto as Exhibit B is a true and correct copy of the field notes of the official Government survey of Township 4 South, Range 2 West, M.D.B. & M., dated February 14, 1894?

The party to whom the request is sent has a specified time to admit or deny the genuineness of each of the documents, usually 30 days.[6] If he admits the genuineness of a particular document, there will be no need to establish the document's authenticity at trial. If the genuineness is denied, and the document is shown at trial to be authentic, the party who denied the authenticity may be held liable for the cost of establishing it, even though he may have won the lawsuit. These costs can be substantial, particularly when proving the authenticity of a document requires producing a witness who lives a great distance from the site of the trial.

The request for admissions can also be used to establish matters of *fact*. Suppose one issue in a lawsuit is the correct location of a section corner, another is the date of death of a certain man, and another is the pre-statehood status of the State of California. An attorney seeking to streamline his case presentation might propound requests for admission in the following form:

> *Request for Admission of Fact No. 1.* Do you admit that the correct location of the northwest corner of section 1, T. 1 S., R. 24 W., Gila and Salt River Meridian, as set by United States Deputy Surveyor John A. Barry in 1902, is as shown on Exhibit A attached hereto?

---

5. *See, e.g.,* Cal. Civ. Proc. Code § 2033(a).
6. *See, e.g.,* Cal. Civ. Proc. Code § 2033(a).

[Exhibit A is a plat of a survey conducted by the surveyor for the pro-
pounding party, showing his placement of the corner in question.]

Request for Admission of Fact No. 2. Do you admit that Gideon
Lightfoot, whom you assert to be a predecessor in interest of Plaintiff,
died on August 4, 1912?

Request for Admission of Fact No. 3. Do you admit that prior to
its admission to statehood on September 9, 1850, the State of California had never enjoyed the status of a territory of the United States of
America?

If a request for admission is admitted, there will be no need to establish the fact at trial;
the attorney will instead read the admission to the jury or judge. On the other hand, if the request for admission of fact is denied, the truth or falsity of the fact in question will be resolved
at trial based on the evidence presented. And, as mentioned above, if the fact is established at
trial, the denying party may be charged with the cost of proving the fact. This economic incen-
tive naturally tends to cause a party to take a request for admission seriously.

D.   REQUESTS FOR PRODUCTION OF DOCUMENTS

Documents, whether on paper or computer memory, form the heart of today's commercial world and, hence, its law cases. One of the most widely used discovery devices thus is
the "request for production of documents."[7] It is precisely what its name suggests. It is a request that the other side produce whatever documents it has that pertain to specified questions. As in the case of interrogatories and depositions, it is not grounds for objection that the
documents contain hearsay, or opinion testimony, or are not relevant for trial purposes. They
must be produced except to the extent that they are subject to valid privileges or cannot meet
the extraordinarily broad discovery concept of relevance.

Requests for production of documents are often made when a witness is called for a deposition. If the witness is a party, the request is just that, made when the witness is given notice of the deposition. If the witness is not a party, the request is made by transmogrifying the
ordinary deposition subpoena into a "subpoena duces tecum," meaning, loosely, a subpoena
("under penalty") under which you must "bring with you" certain things. Those things are
the documents specified in the subpoena.

E.   INSPECTION OF LAND

Another discovery device that is frequently employed in land cases is the request to inspect property. In many states, a land surveyor, for example, does not require the permission
of a landowner to enter his property when the surveyor is conducting a survey.[8] In jurisdictions where a surveyor or other expert does not enjoy this privilege, it will be necessary for
the attorney to request that the surveyor be allowed onto the property. If the request is de-

---

7. Fed. R. Civ. P. 34; Cal. Civ. Proc. Code § 2031.
8. *See, e.g.,* Cal. Civ. Code § 846.5.

nied, the court will invariably order that the surveyor be allowed to enter the property, unless the request is clearly frivolous. In any event, this discovery device is useful to allow inspection by a forester, soils engineer, hydrologist, or other expert retained by the attorney.

Something further should be mentioned of the permitted scope of inquiries made through the discovery process. It is first important to note that what is "relevant," and thus properly inquired into, is much broader in discovery than what is relevant at trial. At trial, evidence and testimony must be "relevant" (see Chapter 1), which for *that* purpose is defined as tending to prove or disprove a fact in issue. During discovery, on the other hand, a matter is "relevant" if it pertains "to the subject matter of the action," or is "reasonably calculated to lead to the discovery of admissible evidence."[9] Thus the matter inquired into during discovery need not be relevant to "the issues," so long as it is relevant to the "subject matter," or is reasonably designed to lead the inquiring party to evidence which he could introduce at trial.

The "subject matter" formulation is somewhat vague, but an example may demonstrate its meaning. Consider a case in which it is disputed whether a corporate officer had authority to execute a deed. An interrogatory asking the date of the company's incorporation may be said to be within the "subject matter" of the action. An interrogatory asking the names, on the other hand, of all company personnel who were in New York on a given day might not be relevant to the subject matter of the case. But in an antitrust case, for example, if one issue is whether a company employee met with the president of a competitor in New York that day, the latter question may well be considered "relevant to the subject matter," and certainly reasonably calculated to lead to the discovery of admissible evidence, and thus be within the purview of discovery.

As mentioned, it is permissible in discovery to elicit hearsay evidence, whether there is an applicable exception or not.[10] To object to an interrogatory or deposition question on the ground that it asks for hearsay (and further, to refuse to answer it for that reason) is to invite a judge to impose "sanctions"—a monetary penalty—as well as to award your opponent his costs and attorneys fees in securing an answer. Again though, all *privileges* that are recognized for trial provide grounds for both objecting and declining to respond to a discovery inquiry.

---

9.   *See, e.g.,* Cal. Civ. Proc. Code § 2017(a); Fed. R. Civ. P. 26(b)(1).

10.   Cal. Civ. Proc. Code § 2017(a), and former. Cal. Civ. Proc. Code §§ 2016(b); 2030(c); Fed. R. Civ. P. 26(b).

Chapter 12

Trial

When you ask God to send you trials, you may be sure your prayer will be granted.

—Leon Bloy

Bloy was not referring to the pleas of the overburdened trial lawyer, but he might have been. For to the trial lawyer (not "litigator" but "trial lawyer"), trials are all too often an embarrassment of riches. They come in bunches, with no respite in between. And they are often scheduled to begin the same day as others the lawyer must try. "Trying" is the resulting modifier.

The trial of a case, it can be said with more force than about any other subject in this book, simply cannot be treated in a chapter. The most that can be tried is an outline of the steps that typically occur in a trial, and of some of the principles that govern its conduct. The problems of omission mentioned in the introduction, as well as of the author's prejudices about what are matters of importance or interest, are more apparent here than in other chapters.

The first procedures to be discussed are actually pretrial events, but they are more coherently treated here in connection with the trial itself.

A. SETTLEMENT CONFERENCES

The parties to a lawsuit may at any time, of course, pursue settlement negotiations with each other in an effort to resolve the controversy short of trial. The law, for reasons as obvious as the finite number of courts and judges available to hear cases, encourages out-of-court settlements, and indeed the rules of discovery themselves have been promoted in part because they tend to facilitate settlements.

Virtually all courts in the United States, including appellate courts at both the state and federal level, have advanced the policy favoring settlement an additional rung by establishing a formal procedure for court-sponsored settlement conferences among parties. The former procedure in California state courts provided for the clerk of the court to send an "invitation" to all parties to a case that had been placed on the "civil active list" (which generally occurs when all parties have appeared and all necessary pleadings have been filed) a specified length of time. If any of the parties accepts the court's invitation, then the matter is calendared for a settlement conference before a judge. The rule is obviously designed to permit the maximum possible flexibility in seeking to attain a settlement: "Settlement conferences shall be held informally before a judge at a time and place provided by the presiding judge. . . . The conference may be continued from time to time by the judge. . . . If the case is not settled at such conference, no reference shall thereafter be made to any settlement discussion had under this rule, except in subsequent settlement proceedings. The settlement procedure provided in this rule is not intended to be exclusive, and local settlement procedures after the completion of pretrial proceedings are expressly authorized if consistent with these rules."[1]

The current California rule, adopted in 1995, provides for a "mandatory settlement conference" in all long-cause cases. A case is considered to be a "long cause" matter if its estimated trial time is more than 5 hours; otherwise it is a "short cause."[2] The Rules of Court require that the parties to a settlement conference prepare a settlement-conference statement containing a good-faith demand or offer of settlement. Moreover, they require that "persons with full authority to settle the case shall personally attend the conference."[3]

The reason for the rule requiring the personal attendance of persons having authority to settle the case should be plain: Without the personal presence of that person, the settlement-conference judge can fall victim to the Mutt-and-Jeff game. The lawyer for a party can claim that he or she would truly like to settle, but some unseen presence far away, having the authority, won't agree. Good settlement-conference judges, like all good arbitrators and mediators, need to have the person with authority personally present. The art of forging a settlement is largely an interpersonal art, consisting of a roiling cauldron of flattery, cajolery, gentle suasion, and old-fashioned Lyndon Johnson-style bullying. If the person with authority is safely ensconced far away, the settlement-conference judge is rendered that much helpless.

Since the first edition of this book, "ADR" has been in the ascendancy. "ADR" is the acronym for "alternative dispute resolution." "Alternative" is itself shorthand for an alternative to litigation. "ADR" comprises mediation, arbitration, and the new "rent-a-judge" phenomenon. Some of the best judges in California, to take just that state, have taken early retirement in order to join the companies that have been formed to provide these services. I would be less than candid if I did not add that the rise of these services is a direct response to (a) the perceived inefficiency of the modern-day court system and (b) the almost universally deplorable state of manners of the litigation bar.

Similar provisions for court-sponsored settlement conferences before trial are found in the local rules of virtually every federal court in the country. Typical provisions are that any party may request a settlement conference under the auspices of the court, and that the judge assigned to preside at the trial of the case shall generally not preside over the settlement conference.[4]

---

1. Former Cal. R. Ct. 207.5 (repealed 1985).
2. Cal. R. Ct. 222, 216.
3. Cal. R. Ct. 222 (c).
4. *See, e.g,* ADR L.R. 7-1 and 7-5, N.D. Cal.; Local Civil Rule 23, C.D. Cal.; and L.R. 16-270, E.D. Cal.

B.   THE PRETRIAL CONFERENCE AND THE PRETRIAL CONFERENCE ORDER

As with many of the modern reforms in the rules of civil procedure, the practice of "pretrying" cases is designed to facilitate the actual trial of the case. Typical of the kinds of matters that are considered in any pretrial conference, whether in state or federal court, is the enumeration contained in Rule 16 of the Federal Rules of Civil Procedure, which provides in subsection (a):

> In any action, the court may in its discretion direct the attorneys for the parties and any unrepresented parties to appear before it for a conference or conferences before trial for such purposes as
> (1)   expediting the disposition of the action;
> (2)   establishing early and continuing control so that the case will not be protracted because of lack of management;
> (3)   discouraging wasteful pretrial activities;
> (4)   improving the quality of the trial through more thorough preparation; and
> (5)   facilitating the settlement of the case.

As veteran trial lawyers can attest, an experienced and firm judge can do much during the course of a pretrial conference to reduce the time required to try a case. (Such a judge can also exert a potent influence on the parties during a settlement conference to achieve a compromise of their dispute.) In complicated cases, the "expediting the disposition of the action" referred to in subparagraph (1) of Federal Rule 16 may include virtually anything that will facilitate the trial of the cause. The court may order all discovery completed by an unusually early date, the filing of memoranda of law on certain critical legal issues, the identification and lodging with the court of all exhibits that will be offered as evidence by the parties, and so on. Provisions similar to Federal Rule 16 are found in the rules of procedure of the various state courts.[5]

---

5.   The reader may be confused at the several references to "local rules of court," "Federal Rules of Civil Procedure," and so forth. In general, it is correct to say that in the federal judicial system as well as in most state-court systems, there are three sources of rules governing the conduct of civil actions. There are, first, the statutes of the Congress or state legislature which set forth the fundamental rules of civil litigation; second, more detailed rules, usually provided by the judiciary itself, applicable to all federal courts, or to all trial courts within the state; and third, special rules of each court. Thus, in the federal system, the acts of Congress providing for the conduct of civil litigation are set forth in the main in Title 28 of the United States Code. Applicable to all district courts of the United States are the Federal Rules of Civil Procedure adopted pursuant to Section 2 of the Act of June 19, 1934 (48 Stat. 1064). Finally, there are the local rules of each federal district court in the United States. (Not to be mentioned are the peculiar rules of each judge of the courts.) Many states have a similar three-tiered structure. In California, for example, the statutes of the legislature providing the fundamental rules of civil procedure are set forth in that state's Code of Civil Procedure; more specific rules applicable to each trial court in the state are set forth in the California Rules of Court; and finally, virtually every trial court in the state, such as the Superior Court for the City and County of San Francisco, has adopted local rules of court.

The product of the pretrial conference is the pretrial order, which is often the cardinal document in the conduct of a lawsuit that goes to trial. In addition to specifying such matters as limits on the number of experts that may be called by either side, it ordinarily serves to (a) supersede the pleadings and (b) restate the issues in controversy. This is a point of great consequence. In the chapter on the pleading stage, it was noted that the issues in a case were framed or "joined" by the operation of the admissions and denials in an answer (as well as by its "affirmative defenses"). That is, all matters in the complaint that are denied in the answer are deemed "joined" and subject to determination by the trial of fact at trial. And all affirmative defenses, while they need not be expressly denied in a written document by the plaintiff, are similarly considered denied and subject to trial. The pretrial order supersedes this earlier framing of the issues and sets forth specifically which questions remain to be tried. Thus, the importance of an attorney's being thoroughly prepared for this conference cannot be overstressed. In the language of Rule 16 of the Federal Rules, the pretrial "order shall control the subsequent course of the action unless modified by a subsequent order."

C.   IN LIMINE MOTIONS

At the time set for trial, the parties may seek rulings in advance from the judge on questions of the admissibility of evidence or testimony. One party may know, for example, that his opponent will seek to introduce into evidence a plat of a private survey, and he may wish that its admissibility be determined *before* it is sought to be introduced at trial. Once it is sought to be introduced—when it is alluded to, for example, or when a witness has answered a question by giving impermissible hearsay in response—the opponent has the proverbial problem of unringing the bell. Such motions in advance are made almost exclusively in jury trials; and because they are made "at the threshold" of the trial, or in Latin "*in limine,*" they are referred to commonly as "*in limine* motions."

In both state and federal procedure, the motion *in limine* is a nonstatutory proceeding that finds its authority in the inherent power of the court to control its process. This power may be found in statutes, for example, that authorize the court to provide for the orderly conduct of proceedings before it, to control its process and orders, and to exclude irrelevant evidence.[6] Motions *in limine* in the author's experience are limited to jury trials, although there seems to be no authority requiring that to be the case.[7] The reason its use appears to be limited to jury trials is, of course, that in a court-tried case, the judge both tries the facts and applies the law; he is going to hear the evidence either way. On the other hand, a jury may be prejudiced by exposure to evidence a judge subsequently rules is inadmissible; an *in limine* motion helps prevent undue prejudice of the jury. The underlying rationale for the use of the motion was best expressed in a 1962 Texas decision:

---

6. *See, e.g.,* Cal. Civ. Proc. Code §§ 128(3), 128(8); and Cal. Evid. Code § 350. As to the use of the motion *in limine* in federal courts, Wright and Graham suggest that Federal Rule of Evidence 103 (a)(1), which requires that an objection to evidence be "timely," constitutes inherent authorization for the motion *in limine,* and for an order on it. Wright & Graham, Federal Practice and Procedure: Evidence § 5037 (1977). *And see* Cal. Evid. Code §§ 353, 354.

7. *See, e.g.,* California Judges' Benchbook, Civil Trials § 3.12, at 73 (1981).

It is the prejudicial effect of the questions asked or statements made in connection with the offer of the evidence, not the prejudicial effect of the evidence itself, which a motion *in limine* is intended to reach."[8]

As the California Judges' Benchbook[9] counsels, while such motions are normally made before the Jury is selected, there is no compelling reason why they must be, and the court may entertain such motions out of the presence of the jury at any time during trial within its discretion.

Violation of an order entered following a motion *in limine* is, like a violation of any court order, a contempt of court.[10] In addition, it may constitute ground for declaring a mistrial or for granting a new trial.[11]

D. PRETRIAL BRIEFING AND JURY INSTRUCTIONS

During the pretrial conference the judge may order that the parties brief some or all of the legal questions applicable to the case and specify the dates on which such briefs or memoranda are to be filed with the court. Typically, such briefs are to be filed on the day set for the beginning of trial. Even in the absence of an order requiring counsel to file a pretrial brief, the lawyer may deem it advisable, particularly if he feels the court's appreciation of subtle points of law is essential to its understanding of his case, whether the case is to be tried by the court or by a jury.

In jury cases, such legal questions may also be briefed incidentally in the course of proferring jury instructions to the trial judge. The practice in federal courts is for the attorneys to submit to the court the instructions they wish the jury to hear *after* the evidence and testimony have been received. On the other hand, it is customary or mandatory in many state courts that proposed jury instructions be submitted in writing to the court *before* the beginning of trial. California Code of Civil Procedure section 607(a) provides, for example, that an attorney must deliver to the trial judge and serve on opposing counsel "before the first witness is sworn . . . all proposed instructions to the jury covering the law as disclosed by the pleadings." It is customary in such courts that other related forms, such as the form on which the jury is to enter its verdict, be submitted at this time as well. In these courts, the trial judge has discretion to refuse an instruction that is submitted in violation of the rule requiring the submission of instructions before trial.[12] Even so, some judges in these jurisdictions prefer to permit counsel to submit their instructions at the close of the plaintiff's case. Their rationale is that if counsel were required to submit the instructions at the beginning of trial, they will submit many more than necessary because neither party knows which issues disclosed by the pleadings will actually be developed at trial. These judges find that the court's time is more

---

8. *Bridges v. Richardson,* 354 S.W.2d 366, 367 (Tex. 1962).

9. California Judges' Benchbook, *supra* note 5.

10. *See, e.g.,* Cal. Civ. Proc. Code § 1209(5); *Charbonneau v. Superior Court,* 42 Cal.App.3d 505, 116 Cal.Rptr. 153 (1974).

11. *See, e.g.,* California Judges' Benchbook, *supra* note 5 at 332–333, 545–546.

12. *See, e.g., Wilson v. Gilbert,* 25 Cal.App.3d 607, 102 Cal.Rptr. 31 (1972).

expeditiously handled if they permit this delay and encourage the attorneys to consult with each other to try to come to agreement on as many of the instructions as the jury will receive as possible.

The rule, as mentioned, is different in federal courts. Rule 51 of the Federal Rules of Civil Procedure provides the following:

> At the close of the evidence or at such earlier time during the trial as the court reasonably directs, any party may file written requests that the court instruct the jury on the law as set forth in the requests. The court shall inform counsel of its proposed action upon the requests prior to their arguments to the jury. The court, at its election, may instruct the jury before or after argument, or both. No party may assign as error the giving or the failure to give an instruction unless that party objects thereto before the jury retires to consider its verdict, stating distinctly the matter objected to and the grounds of the objection. Opportunity shall be given to make the objection out of the hearing of the jury.

E. WHAT LAW?

The "choice of law" is a subject that has with surprising frequency been before the United States Supreme Court in the context of land cases in recent years. In a trial court, the choice-of-law question may be raised in an *in limine* motion, during argument in a case respecting jury instructions, or in pretrial briefs. For want of any markedly more appropriate location, it will be treated here.

The question, essentially, is what body of law is to provide the rule of decision on an issue. In property cases, the law of the state in which the property is situated is ordinarily the applicable law.[13] The laws of the various states, however, and the body of so-called "federal common law" often differ in their treatment of title and boundary issues. As an example, California has a peculiar rule in boundary law that, in certain cases, accretion to waterfront property caused by the activities of man does not result in an addition to the upland estate.[14] In California such "artificial" accretions, when they occur because of the erection of a levee, for example, leave the boundary in its "last natural condition"[15]—at least in the long-held view of state officials. The federal rule, on the other hand, is the more traditional common-law rule. It states that accretion—that is, the gradual and imperceptible addition to the soil on the bank or shore of the waterbody—becomes an addition to the upland estate

---

13. *See, e.g., Packer v. Bird,* 137 U.S. 661, 669 (1891); *cf. Borax, Consolidated v. Los Angeles,* 296 U.S. 10, 22 (1935); *see also Hughes v. Washington,* 389 U.S. 290, 295 (Stewart, J., concurring) (1967).
14. California does recognize that accretion naturally formed by the deposition of sediments on the bank or shore of the water course inures to the upland owner and, in effect, causes his boundary to move waterward.
15. *See, e.g.,* Cal. Civ. Code § 1014; *Carpenter v. City of Santa Monica,* 63 Cal.App.2d 772, 787 (1944); *United States v. Aranson,* 696 F.2d 654 (9th Cir. 1983); *State v. Superior Court (Lovelace),* 11 Cal. 4th 50 (1995).

regardless of whether it is caused solely by natural causes or has been influenced by an artificial agency.[16]

Obviously, then, it may be of great consequence to the litigants in an accretion case which body of law will apply to the facts of their dispute. Principally, these questions have arisen in cases of property abutting the ocean or bounded by rivers that form, or once formed, interstate boundaries. They have arisen as well in cases of the boundaries of Indian reservations. A brief discussion of some of the recent United States Supreme Court decisions in this field may be of interest to the reader.

While many such choice-of-law cases were decided in the nineteenth century, it is convenient to begin with *Borax Consolidated v. Los Angeles*, 296 U.S. 10 (1935), which held that the extent and validity of a federal grant (in that case a patent to a portion of Mormon Island in San Pedro Bay) was a question to be resolved by federal law, not state law. In 1967, the Supreme Court again confronted a similar situation in *Hughes v. Washington*, 389 U.S. 290 (1967). There the court wrote:

> The question for decision is whether federal or state law controls
> the ownership of land, called accretion, gradually deposited by the
> ocean on adjoining upland conveyed by the United States prior to
> [Washington's] statehood. . . . We hold that this question is governed
> by federal, not state, law and that under federal law Mrs. Hughes,
> who traces her title to a federal grant prior to statehood, is the owner
> of these accretions.[17]

The issue was again presented to the United States Supreme Court a mere 6 years later in *Bonelli Cattle Co. v. Arizona*, 414 U.S. 313 (1973). In that case, the dispute was between a private landowner, the Bonelli Cattle Co., and the State of Arizona over the ownership of lands that "re-emerged" from the bed of the Colorado River following a rechannelization project that lowered the level of the river. As in *Hughes,* the State of Arizona had a rule which provided that such "accreted" lands (to be precise, in *Bonelli* the lands were relicted, not accreted) did not belong to the upland owner, but became the property of the state. The result, of course, would be different under the federal rule mentioned above. The Court's holding in *Bonelli* was 4 years later overruled by the Supreme Court, which characterized the *Bonelli* holding as follows:

> We held [in *Bonelli*] that federal common law should govern in
> deciding whether a State retained title to lands which had re-emerged
> from the bed of a navigable stream, relying in part on *Borax, Ltd. v.
> Los Angeles*, 296 U.S. 10 (1935).
>
> We went on to discuss the nature of the sovereign's [the State of
> Arizona's] interest in the riverbed, which we found to lie in the protection of navigation, fisheries, and similar purposes. We held that under federal common law, as we construed it in that case, Arizona's

---

16.   The leading case on this subject is *County of St. Clair v. Lovingston,* 90 U.S. (23 Wall.) 46 (1874).

17.   389 U.S. at 290–291.

sovereign interest in the re-emerged land was not sufficient to enable it to retain title. We found that the principle governing title to lands which have been formed by accretion, rather than that which governs title where there has been an avulsive change in the channel of the river, to be applicable. We chose the former because it would both ensure the riparian owner access to the water's edge and prevent the State from receiving a windfall. We therefore decided that Bonelli, as riparian owner, was entitled to the land in question.[18]

That description of *Bonelli* was written by Justice Rehnquist in the case that overruled *Bonelli, State Land Board v. Corvallis Sand & Gravel Co.*, 429 U.S. 363 (1977). The *Corvallis* case involved the ownership of certain lands underlying the Willamette River, a navigable river which the Court pointedly noted is not an interstate-boundary river. A protracted discussion of the opinion would be inappropriate here, but suffice it to say that the Court felt clearly that its *Bonelli* decision of only 4 years earlier was wrong, and it held that state law, not federal, governed the title to the accreted lands.

The next decision in this line of cases was *Wilson v. Omaha Indian Tribe*, 442 U.S. 653 (1979), which, as the title of the case implies, added more ingredients to the kettle. As in the other cases, a state's sovereign title to the bed of a navigable body of water was in issue. Also, *Wilson* also involved a stream that formerly had constituted the boundary between two states, Iowa and Nebraska. (The same considerations were historically present in the *Bonelli* case, which concerned the Colorado River. The Colorado River had also, by interstate compact, ceased to serve as the interstate boundary between Arizona and California.) In addition, though, one of the competing claimants in *Wilson* was an Indian tribe. In brief, Iowa was admitted to statehood in 1846, its western boundary being described as the center of the main channel of the Missouri River. The Omaha Indian Tribe was ceded lands on the Nebraska side of the river in 1854; those lands were bounded on the east by the main channel of the river. While the center of the main channel of the river had formed the common political boundary between Iowa and Nebraska initially, its many shifts over the years caused the two states to fix the boundary in location by interstate compact in 1943.[19] The State of Iowa (as claimant to the bed of the east half of the Missouri River) and several individuals argued that the past movements of the Missouri River had eroded away part of the Omaha reservation, with soil accreting to the Iowa side of the river, vesting title in them as riparian landowners. Given this convoluted state of affairs, the Court stated that the issue was "whether federal or state law determines whether the critical changes in the course of the Missouri River in this case were accretive or avulsive."[20]

If one has studied philosophers such as Aquinas (who pondered the number of angels that could be assembled upon the head of a pin) and Immanuel Kant (who enunciated largely incomprehensible principles of logic), then one will presumably have little difficulty with the reasoning of the Supreme Court's decision in *Omaha*. Notwithstanding the urgings of some

---

18. This characterization of the *Bonelli* case was contained in *State Land Bd. v. Corvallis Sand & Gravel Co.*, 429 U.S. 363, 368–370 (1977), discussed in the next paragraph.

19. *Nebraska v. Iowa*, 406 U.S. 117, 118–119 (1972).

20. 442 U.S. at 658.

30-odd states that the Court hold state law applicable, the Court held that federal law was the appropriate choice under the circumstances. That having been decided, the Court thereupon ruled that, under the circumstances, federal law ought to "borrow" state law in the case.[21] So it did; but the state whose law was borrowed was Nebraska, not Iowa, notwithstanding that the land lay in Iowa!

The Supreme Court once again dealt with choice of law 3 years later in a case of a title claim by the State of California to accreted oceanfront lands on the California coast within (or adjacent to, as one perceived the case) a Coast Guard reservation. The application of California's artificial-accretion doctrine would have vested title in the state, because the accretion was caused by the construction of training jetties at the mouth of Humboldt Bay. Application of the federal rule, as enunciated in *County of St. Clair, supra,* would have secured the title to the United States. California's endeavor was repelled in the end, the Supreme Court holding that the federal government's needed security of title, as well as such precedents as the *Hughes* decision, mandated the application of federal law to the controversy.[22]

It is hoped that the reader will excuse this protracted and somewhat tongue-in-cheek digression from the ostensible subject of this chapter, but ideally he will see that there are many more things in the law, as "in heaven and earth . . . than are dreamt of in your philosophy."

F.   SELECTION OF THE JURY

While quiet-title actions are almost always tried without a jury, ejectment and trespass are normally tried before juries. (The reasons that title actions are normally tried to the judge are explained below.) In jury cases the next principal step in the process of trial is the selection of the jury.

It had long been assumed that a jury, whether in a civil case or criminal case, must consist of 12 persons, and the right to have a 12-member jury is guaranteed in some states by either statute or the state's constitution. Article I, section 16 of the California Constitution, for example, provides a right to a 12-person jury unless the parties stipulate to a lesser number. In California and several other states, the agreement of three-fourths of the jurors is necessary for a verdict.[23] In the federal system, however, the size of the jury is a matter of some perplexity. Rule 48 of the Federal Rules of Civil Procedure formerly provided that parties could stipulate to a jury of fewer than 12 persons or that the verdict or finding of a stated majority of the jurors may be taken as the verdict or finding of the whole. Federal Rule 48 thus seemed to imply that there was a right to a 12-person jury, an implication deriving from the long-maintained assumption that the Constitution provides a right to a jury of 12. In 1970, however, the United States Supreme Court held in *Williams v. Florida*[24] that the right to a jury trial in a criminal case (protected by the Sixth Amendment to the Constitution and binding

---

21.   442 U.S. at 672–674.

22.   *California ex rel. State Lands Comm. v. United States,* 457 U.S. 273 (1982). (As it happens, both protagonists in that case, Deputy Solicitor General Louis Claiborne, and California Deputy Attorney General Bruce Flushman, are now law partners of mine.)

23.   *See* Cal. Civ. Proc. Code § 194.

24.   *Williams v. Florida,* 399 U.S. 78 (1970). *Ballew v. Georgia,* 435 U.S. 223 (1978) held that six was the minimum size of a jury.

on the states through the Fourteenth Amendment) is not a right to a jury of 12, and that a six-man jury in a *criminal* case satisfied the constitutional requirement. Three years later, in *Colgrove v. Battin,* 413 U.S. 149 (1973) the Court ruled that a six-person jury in civil cases was unconstitutional, under the Seventh Amendment. Following the *Williams* and *Colgrove* decisions many federal district courts adopted local rules providing for a jury of fewer than 12 in some or all *civil* cases. The number adopted by the district courts was usually six,[25] a number which has been endorsed by the Judicial Conference of the United States. Rule 48 was amended in 1991 to provide that juries in civil cases must consist of not fewer than six and not more than 12 jurors.

The process of jury selection begins with a list of potential jurors, sometimes called the "panel," from which the names of 6, 8, or 12 people—whatever the total number of jurors for the case will be—are drawn. These people are then seated in the jury box and the process of examining them, "voir dire" (roughly, "to see to say") begins. The mechanics of this process vary considerably from court to court, but its ostensible purpose in all cases is to determine whether there exists any reason for disqualifying one of the potential jurors. Reasons for disqualification are fairly self-evident, such as a relationship to one of the parties or counsel, a prejudice in favor or against one of the parties, and so on.[26] In federal courts, it is more typical for the judge to do most, if not all, of the questioning of the prospective jurors. Rule 47(a) of the Federal Rules of Civil Procedure provides that either the judge or the attorneys may question the prospective jurors on voir dire, but the local rules of the district courts in more cases than not provide that the trial judge is to conduct the examination. The attorneys of course may submit questions that they wish the judge to ask the prospective jurors. In many state courts, greater latitude is given to the attorneys to examine the prospective jurors, and the judge confines himself to more perfunctory matters such as whether any of the jurors are related or acquainted with any of the parties or the attorneys.[27]

There is no limit on the number of prospective jurors who may be successfully challenged "for cause." By this is meant that the juror is demonstrated to have some bias or other disqualification to the satisfaction of the judge. It is of course a delicate matter for an attorney to decide to challenge a juror for cause, because ordinarily he must make his challenge in the presence of that juror. If he is unsuccessful in demonstrating to the court's satisfaction that the juror has some disqualification, the attorney will have to abide that juror during the course of trial and perhaps endure the effects of any offense he may have caused that juror.

In addition to challenges for cause, each side in a jury trial is given a specified number of challenges without cause, commonly called "peremptory challenges." These challenges may be exercised by an attorney without any stated reasons and are typically used on prospective jurors whom the attorney feels will be ill-disposed toward his case, but whose bias he feels he cannot demonstrate. Often, a juror who has been unsuccessfully challenged for cause will be peremptorily challenged by the same attorney, for obvious reasons. The number of peremptory challenges allotted to each side varies from jurisdiction to jurisdiction;

---

25.   *See e.g.,* Local Civil Rule 13.1, C.D. Cal.
26.   *See, e.g.,* Cal. Civ. Proc. Code § 225.
27.   *See, e.g.,* Cal. R. Ct. 228, which requires the trial judge to permit counsel to examine prospective jurors, "within reasonable limits."

they can be as many as six per side in civil cases.[28] In the federal judiciary, each party is typically allowed three peremptory challenges.[29] Where there is more than one party per side, the rules respecting the number and allotment of peremptory challenges get somewhat complex.

In the federal system, a unanimous jury verdict is required in civil as well as criminal cases.[30] In other jurisdictions, something less than unanimity is frequently permitted.[31]

It was mentioned earlier that quiet-title actions are not normally tried before a jury. In England in the eighteenth century (when the American Revolution occurred), as well as earlier, such actions were tried in courts of equity, and not in the common-law courts. The equity or Chancery courts (they derived from the ecclesiastical courts), unlike the common-law courts, did not provide for the right to a jury. The text of the Seventh Amendment to the United States Constitution, which is commonly thought to guarantee an absolute right to a jury trial in civil cases, actually provides that "[i]n suits *at common law* . . . the right of trial by jury shall be preserved . . . ." (Emphasis added.) Thus in actions to quiet title, as well as other actions such as injunction and specific performance of contract, which were historically tried in the courts of equity, there is no right to a trial by jury. Comparable provisions are found in virtually all of the constitutions of the various states.

While quiet-title actions are normally equitable in character, and even though they may remain nominally actions to quiet title, if they ask for return of possession of the land they become more akin to the old legal form of action called "ejectment." Because ejectment is an action that was tried before the *common-law* courts of England (i.e., it was "legal" in character as opposed to "equitable"), the right to a jury attaches.[32]

If the action raises some legal (e.g., possession) and some equitable (e.g., title) issues, a party is entitled to a jury trial on the legal issues. In such a case, although in theory there is only a single trial and a single judgment and appeal, as a practical matter there may be two separate and distinct proceedings. Sometimes the entire trial will be by jury, but the court will treat the verdict on the equitable issues as advisory only.[33] In condemnation actions, there is no right to a jury trial on the issue of whether the defendant, for example, has a prescriptive easement over adjoining land. This issue was in the nature of quiet title, and the state and federal constitutional guarantees of a jury trial in civil cases extends only to the question of just compensation in eminent-domain proceedings.[34] Nevertheless, if the title issue was submitted to the jury in a condemnation case without objection, the fact that it was decided by the jury is not grounds for reversing the judgment.[35]

---

28.   *See, e.g.,* Cal. Civ. Proc. Code § 231 (c).

29.   28 U.S.C. § 1870.

30.   *See, e.g., American Publishing Co. v. Fisher,* 166 U.S. 464, 468 (1897). ("Now unanimity was one of the peculiar and essential features of trial by jury at common law").

31.   *See, e.g.,* Article I, § 16 of the California Constitution.

32.   *See Thomson v. Thomson,* 7 Cal.2d 671, 681, 62 P.2d 358 (1936); 25 Cal. L. Rev. 565 (1937); 65 Harv. L. Rev. 458 (1952). A simple action to quiet title, with possession of the property not in issue, is purely equitable. *Santa Anna M. & I. Co. v. Kinslow,* 30 Cal.App.2d 107, 110, 85 P.2d 899 (1938); *Campbell v. Rustigian,* 60 Cal.App.2d 500, 140 P.2d 998 (1943). *Cf. Southern Pac. Land Co. v. Dickerson,* 188 Cal. 113, 116, 204 P. 576 (1922).

33.   *See, e.g., Hughes v. Dunlap,* 91 Cal. 385, 388, 27 P. 642 (1891).

34.   *Pacific Gas & E. Co. v. Peterson,* 270 Cal.App.2d 434, 438 (1969).

35.   *City of Oakland v. Wheeler,* 34 Cal.App. 442, 455–56 (1917).

G.   OPENING STATEMENTS, AND THE PRESENTATION OF EVIDENCE AND TESTIMONY

Following jury selection, the trial of the case begins with the giving of opening statements by the attorneys. In nonjury cases where the court has extensive pretrial briefing, the judge may order the parties to dispense with opening statements and simply commence the presentation of their cases. In jury trials, the plaintiff's attorney gives his opening statement first, which can be but a summary of the evidence that the plaintiff intends to introduce at trial. Opening statements are extremely critical in jury cases, because studies have shown that the first impressions formed by jurors when they hear the opening statements of counsel correspond closely to how they vote during deliberations.

The opening statement is not simply an opportunity for the plaintiff's attorney to create a favorable impression with the jury, however. He is required to describe the evidence and testimony he will seek to introduce on every issue on which he bears the burden of production. If he neglects to describe sufficient evidence on every such issue, he risks what is called a "nonsuit." Defendant's counsel, following an opening statement in which the plaintiff's attorney neglects to describe such evidence, may move the court for a nonsuit. If granted, a nonsuit has the effect of ending the case without the defendant even having to make *his* opening statement. (A nonsuit may also be granted following the conclusion of plaintiff's case in chief, if he fails to produce evidence to support a finding in his favor on every issue on which he bears the burden of production. This event is discussed more thoroughly later.)

Following the plaintiff's opening statement, the defendant may make his opening statement, if he wishes. In most jurisdictions the defendant has the alternative of allowing the plaintiff to go on with the presentation of his case and to defer his own opening statement until the plaintiff has concluded his case in chief.

After opening statements, the presentation of evidence and testimony is begun. In almost all cases, the plaintiff opens the case, calling his witnesses, offering into evidence the documents and other tangible exhibits (such as models, soil samples, etc.), and reading into the record other matters he feels necessary to make out his case. These matters may include his opponent's answers to interrogatories, formal admissions, and so on. Following the presentation by the plaintiff of his case, the defendant then makes his opening statement, if he had declined to make it at the opening of trial, and presents his testimony and evidence. After the defendant's case, most courts then allow the plaintiff to put on rebuttal evidence, which is limited in scope to matters raised by the defendant in his case in chief. The plaintiff is not allowed simply to augment his principal case nor to introduce testimony or evidence that he had neglected to present during his case in chief, except with leave of court. Following plaintiff's rebuttal, some courts allow the defendant to introduce evidence on "surrebuttal," which likewise is limited to matters in response to evidence introduced on rebuttal. This sequence of presentation is fairly typical of all courts.[36]

In eminent domain cases, where a government entity (or private company such as a utility) having the power of eminent domain is acquiring land through condemnation proceedings, the order of proof may be reversed. Some courts in these cases make it customary for

---

36.   *See, e.g.,* Cal. Civ. Proc. Code § 607 (jury trials). The order of presentation is the same in nonjury trials. Cal. Civ. Proc. Code § 631.7.

the defendant (i.e., the landowner) to present his case first, rather than the plaintiff (the condemning agency). In California state courts, for example, it used to be the rule that the landowner had the burdens of both persuasion and production on the question of the value of his property and, on that account, would have the ability to make his opening statement first, as well as to open and close final arguments. The ability to open and close final argument— to have the first and last word—is universally considered a major tactical advantage. Today, however, while the landowner, the defendant in a condemnation case, has the burden of going forward with the evidence and thus putting on his case first, neither party bears the burden of persuasion on the issue of the value of the property being acquired.[37]

It was mentioned earlier that should the plaintiff in his opening statement neglect to describe sufficient evidence that he intends to introduce to support a finding in his favor on every issue on which he bears the burden of production, he risks a "nonsuit." The same principle applies following the plaintiff's actual presentation of his case in chief. If the plaintiff neglected to introduce evidence sufficient to support a finding in his favor during his case, the defendant may in this situation also move for a "nonsuit." That nomenclature, although somewhat archaic, is still found in many jurisdictions, such as California.[38] In the federal system, the motion is for "judgment as a matter of law." It may be used in jury cases only, can be made against any party, but may be made only after that party "has been fully heard on an issue."[39] There are a handful of rules peculiar to the trial of title cases that have proved to be pitfalls for the lawyer unfamiliar with them and have given rise to nonsuits. These rules will be discussed briefly a little later in this chapter.

A witness testifying under oath during trial may testify only in response to questions posed to him by the attorneys. The questioning of a witness by the attorney calling him to the stand is called "direct examination." When the opposing attorney or attorneys are questioning the witness, this interrogation is called "cross-examination." Ordinarily the attorney calling the witness may, following cross-examination, have the opportunity to conduct "redirect examination," following which the opposing counsel may have the opportunity to ask questions on "re-cross-examination." As in the case of rebuttal and surrebuttal, cross-examination is limited in scope to matters raised by the attorney who called the witness to the stand in the course of his questions to that witness.[40] Similarly, redirect examination is limited in scope to matters raised on cross-examination, and the same principle applies to re-cross-examination. Should an attorney on cross-examination, for example, wish to inquire of the witness as to matters that were not raised during the witness's testimony on direct examination, he must content himself to call the witness as his own during his case in chief.

Ordinarily, an attorney who has called a witness to the stand may not examine him by the use of "leading questions"; a cross-examiner, however, may.[41] A leading question, while not always an easy thing to detect, is in general one that suggests the answer the examiner would like to hear. Questions introduced with the principal clause, "Isn't it true that . . ." are classic examples of a leading question. A great deal of latitude is ordinarily permitted by the

---

37.   Cal. Civ. Proc. Code § 1260.210.

38.   *See, e.g.,* Cal. Civ. Proc. Code § 581(c); *Castro v. Adams,* 153 Cal. 382 (1908).

39.   Fed. R. Civ. P. 50(a). *see also* 9 Wright and Miller, Federal Practice 388 (1995).

40.   Fed. R. Evid. 611(b).

41.   *See, e.g.,* Cal. Evid. Code § 767.

trial court in permitting an attorney to ask leading questions of his witness about "preliminary matters" or matters that are not reasonably subject to dispute. For example, an attorney who has called his client to the stand will invariably be permitted to ask the question, "You are the plaintiff in this action, is that correct?"; he will not be compelled to ask the nonleading question, "Are you by any chance a party to this action?"[42] There is one principal exception, however, to the rule forbidding the use of leading questions on direct examination. This exception occurs when an attorney calls to the witness stand a "hostile" or adverse witness. An example of such a witness would be the defendant when called to the witness stand by the plaintiff's attorney. Under these circumstances, the attorney is allowed to examine the witness as though the questioning were cross-examination.[43]

It was mentioned that there are rules peculiar to the trying of title cases that need to be carefully borne in mind by the attorneys in preparing for the presentation of their cases. For one, it is well-settled that the plaintiff in a quiet-title lawsuit, or in any other in which the issue of title is raised (such as condemnation or trespass), must recover on the strength of his own title and not simply on the weakness of defendant's title.[44] Stated simply, this rule means that it is not sufficient for the plaintiff to content himself with introducing evidence and testimony that tends to defeat the basis on which the defendant claims title; he must show affirmatively that title is vested in himself.

There are several corollaries to this rule. A plaintiff relying on a "paper title" (i.e., a chain of title from a sovereign grantor), as opposed to possession for the period of limitations, must trace his title to either (1) a government grantor, (2) a grantor in possession at the time of the conveyance to the plaintiff, or (3) a source common to the chains of title of both plaintiff and defendant.[45] In one case, it was felt that the mere introduction of a deed from the plaintiff's immediate grantor was insufficient to establish a prima facie case, and a nonsuit could properly have been granted.[46] As stated in another case:

> Of course, where both parties claim title from a common source, it is not necessary to prove title in the grantor. But that was not the situation here. The plaintiffs' claim, as already stated, was under Ferdinand Smith, while the defendants relied upon a patent from the state of California to one Craig, and a subsequent tax deed to one Tyler, who had deeded to the defendant. There is no need, therefore, to consider whether the tax sale and deed were regular and valid. The plaintiffs having failed to show either title or possession in themselves, they "cannot complain that someone else, also without title, asserts an interest in the land. . . . A defendant in such an action may

---

42.   *See, e.g.,* Fed. R. Evid. 611(c).

43.   *See, e.g.,* Fed. R. Evid. 611(c); Cal. Evid. Code § 776.

44.   *See, e.g., Winter v. McMillan,* 87 Cal. 257 (1890); *Ernie v. Trinity Lutheran Church,* 51 Cal.2d 702, 706 (1959); *Helvey v. Sax,* 38 Cal.2d 21, 23 (1951).

45.   *Rockey v. Vieux,* 179 Cal. 681, 682 (1919); *Helvey v. Sax,* 38 Cal.2d 21 (1951); *People v. Southern Pac. R.R. Co.,* 166 Cal. 627, 629 (1913).

46.   *Coffin v. Odd Fellows Hall Assn.,* 9 Cal.2d 521, 525 (1937); *San Diego Improvement Co. v. Brodie,* 215 Cal. 97, 100 (1932).

always effectually resist a decree against himself, by showing simply that the plaintiff is without title." And, having shown no interest in the land, the plaintiffs are not aggrieved by a judgment declaring someone else to be the owner. [Citations omitted.][47]

Conversely, when the plaintiff has made out a prima facie showing of title, the defendant cannot question the plaintiff's title where he shows no title or interest in himself.[48] But, as indicated by the *Rockey* case above, where the plaintiff is not in possession of the premises, the defendant may resist a decree against himself simply by showing that the plaintiff has no title.[49]

It is also well-settled that the plaintiff in a quiet-title action must be shown to have been vested in title at the commencement of the action in order to maintain his cause.[50]

Generally speaking, when a defendant has answered the plaintiff's complaint by denying the title of plaintiff, he is free to introduce evidence on any theory that would tend to defeat the plaintiff's claim of title.[51] There are, however, peculiar problems in the pleading and proving of the theory of equitable estoppel in title actions:

As a general rule when estoppel is an element of the cause of action it must be specially pleaded. This rule is particularly applicable where a plaintiff relies upon estoppel to avail against a defense which would otherwise appear upon the very face of the complaint. The rule does not apply, however, under the principle that a complaint need not [be] negative or anticipate any defense on the part of the defendant, where estoppel is relied upon by the plaintiff to meet new matter affirmatively pleaded in the answer, or where it is made relevant by a contention raised therein. Moreover, an estoppel does not have to be pleaded where the party in whose favor it exists at the time of preparing his pleading is without knowledge that his claim must ultimately rest on it. [Citations omitted.][52]

Not infrequently, "extrinsic" evidence is needed to construe the meaning of a document of title that is either (a) ambiguous on its face or (b) rendered ambiguous by other facts. The courts vary greatly in the extent to which they permit extrinsic evidence to be introduced for this purpose. Some courts forbid the introduction of evidence to interpret an instrument such as a deed if its meaning is plain "on its face"—that is, if no ambiguity appears from the document itself. Other courts, however, hold that extrinsic evidence is admissible to show that a deed is ambiguous. One leading California case has held:

---

47.   *Rockey v. Vieux, supra* note 43, 179 Cal. at 682–683.
48.   *Pacific States S. & L. Co. v. Warden,* 18 Cal.2d 757 (1941); *McPhail v. Nunes,* 48 Cal.App. 383 (1920).
49.   *Williams v. City of San Pedro, etc. Co.,* 153 Cal. 44 (1908).
50.   *Reed v. Hayward,* 23 Cal.2d 336 (1943).
51.   *See, e.g., Harmon Enterprises, Inc. v. Vroman,* 167 Cal.App.2d 517 (1959).
52.   *Estate of Pieper,* 224 Cal.App.2d 670, 691 (1964); *see also* 3 B. Witkin, California Procedure, *Pleading,* §§ 944–948 (2d ed., 1971).

The test of admissibility of extrinsic evidence to explain the meaning of a written instrument is not whether it appears to the court to be plain and unambiguous on its face, but whether the offered evidence is relevant to prove a meaning to which the language of the instrument is reasonably susceptible.

A rule that would limit the determination of the meaning of a written instrument to its four-corners merely because it seems to the court to be clear and unambiguous, would either deny the relevance of the intention of the parties or presuppose a degree of verbal precision and stability our language has not attained.

* * *

Although extrinsic evidence is not admissible to add to, detract from, or vary the terms of a written contract, these terms must first be determined before it can be decided whether or not extrinsic evidence is being offered for a prohibited purpose. [Citations omitted.][53]

An interesting corollary to the problem of extrinsic evidence to construe an instrument is the rule that documents of conveyance are to be construed according to the laws in force at the time they were executed. Sometimes, the laws in force were those of a different *country* from the one that is sovereign at the time of trial. "Any question as to the title of property, acquired prior to the conquest, is to be determined by reference to the Mexican law at the time the property was acquired. Any question, however, as to restraint or restrictions upon the use of property, acquired prior to the conquest, is to be determined according to American law."[54]

The total body of such rules of pleading and proof applicable to lawsuits concerning the title to property are too numerous for inclusion here. These examples should suffice for the purpose of apprising the reader that such very particular rules exist—and she should beware!

While the attorneys themselves largely make the decisions about the order in which they wish to present their evidence and testimony (an order that is often dictated by the availability of witnesses during the trial), the trial judge nevertheless retains a great deal of discretion in governing that order.[55]

H.   SUMMATION

When all of the testimony has been given and evidence has been submitted, it is the time for the attorneys to deliver their closing arguments to the jury. While their opening statements were confined to mere descriptions of the evidence they hoped to adduce during

---

53.   *Pacific Gas & E. Co. v. Thomas Drayage, Etc. Co.,* 69 Cal.2d 33, 37, 39 (1968). The *Thomas Drayage* case dealt specifically with a written contract, but the principle has been held applicable to deeds and other documents of conveyance as well. *Murphy Slough Assn. v. Avila,* 27 Cal.App.3d 649, 655 (1972); *Ferris v. Emmons,* 214 Cal. 501 (1931).

54.   *Hart v. Gould,* 119 Cal.App.2d 231, 236 (1953).

55.   *See, e.g.,* Fed. R. Evid. 611; Cal. Evid. Code § 320.

the trial, and theoretically at least were devoid of argument, the closing "argument" is ostensibly the first opportunity for the attorneys actually to argue their case, emphasizing the credibility or lack of credibility of witnesses, suggesting inferences that ought to be drawn from certain testimony, and so on. I say "ostensibly" because lawyers habitually seek to "argue" their case before the jury at every opportunity—for example, when objecting to evidence and even when examining a witness. Arguing the case in the presence of the jury at any time prior to the closing statement is, strictly speaking, improper, and this kind of tactic accounts for the oft-heard objection that "Counsel is arguing his case, your honor," or "The question is argumentative." Trial judges in most jury trials have at least one occasion to admonish a lawyer for arguing.

In nonjury cases, which are conducted much less formally, oftentimes the trial judge will dispense with hearing the closing argument and ask the parties instead merely to brief the case in memoranda. In such cases, following whatever argument and briefing the court wishes, there is usually a written decision that explains in greater or lesser detail reasons for the judge's ruling. Following the written notice of the judge's decision and sometimes following the preparation of more formal "findings of fact and conclusions of law," the judgment is signed by the court and entered on its judgment roll.

In jury cases, following the closing arguments of counsel, the judge instructs the jury on questions of law, and then they retire to deliberate and, it is hoped, arrive at a verdict. When they have reached a verdict, it is announced in open court. The attorneys, if they wish, are then allowed to "poll" each individual juror in order to ascertain that in fact there are sufficient votes to make a verdict. Finally, usually within a very brief space of time, the verdict is entered on the court's records as the judgment of the court.[56]

For most purposes, the entry of judgment is the conclusion of proceedings in the trial court. There are several motions that may be made by the parties following the decision of the trial court, such as a motion for a "judgment notwithstanding the verdict" or for a new trial. These proceedings will be dealt with in the next chapter.

---

56.   "When trial by jury has been had, judgment must be entered by the clerk, in conformity to the verdict within 24 hours after the rendition of the verdict, whether or not a motion for judgment notwithstanding the verdict be pending. . . ." Cal. Civ. Proc. Code § 664.

Chapter 13

Post-Trial Proceedings in the Trial and Appellate Courts

"Appeal, v.t. In law, to put the dice into the box for another throw."

—Ambrose Bierce, *The Devil's Dictionary* (Garden City, NY: Doubleday & Company), p. 17

"I'll fight this case to the Supreme Court if I have to!" is the oft-heard defiant vow of the party who loses at trial. (That is, until he receives his attorney's bill for representing him thus far[1]).

As we shall see, though, taking an appeal, to the Supreme Court or elsewhere, is not simply a matter of taking a second bite of the apple. An appeal is not a retrial of the case in higher courts. Only very select questions of law are heard on appeal, and with rare exceptions these do not include a reconsideration of the evidence or an independent determination by the reviewing court of the facts. Also, review by the highest courts—the United States Supreme Court and the state supreme courts, for example—is ordinarily not a matter of right. These courts have almost unrestricted discretion in deciding what cases to hear, and they take only those that present significant questions of law of pervasive interest in the society far beyond what might constitute justice as among the parties to the particular cases. The United States Supreme Court, for example, hears argument in only two percent of the cases it is asked to review, about half the percentage of only 20 years ago.

Before a case may be appealed, however, a number of procedural maneuvers can take place in a trial court following the decision. The first part of this chapter will consider five of

---

1. Among lawyers of the San Francisco Bar the story is told of Garret McEnerney, one of its most distinguished members, who died in 1942; he had litigated a major antitrust suit in the courts for 10 years, only to lose in the United States Supreme Court. When he had finished reviewing the final bill to his client he signed it, paused, and then added the postscript, "Perhaps you think this case might have been lost for less."

these procedural matters: The entering of "findings of fact and conclusions of law" in non-jury cases, applications for stay of execution, determinations whether costs or other litigation expenses are to be awarded either party, motions for judgment "n.o.v." or "notwithstanding the verdict" in jury cases, and motions for new trial. The second part of this chapter will then address the appeal of cases.

A.   POST-TRIAL PROCEEDINGS BEFORE APPEAL

FINDINGS AND CONCLUSIONS IN NONJURY CASES

To ensure a clear "record" on appeal, many jurisdictions have adopted the practice in nonjury cases of requiring the trial court to render a formal statement of reasons for its decision, usually cast in the form of "findings of fact and conclusions of law." The "record" of the trial is the only means by which an appellate court can review a lower court's decision. Thus, to intelligently conduct its review, the appellate court needs the clearest possible statement of the reasons for the decision of the trial court. Rule 52 of the Federal Rules of Civil Procedure provides, for example, that "in all [United States District Court] actions tried upon the facts without a jury or with an advisory jury, the court shall find the facts specially and state separately its conclusions of law thereon. . . . Findings of fact . . . shall not be set aside [on appeal] unless clearly erroneous, and due regard shall be given to the opportunity of the trial court to judge of the credibility of the witnesses."[2]

In a quiet-title action, for example, which being equitable in nature is ordinarily tried without a jury, typical findings of fact might read as follows:

> 1.   The subject property was first surveyed for the purpose of subdivision under the public land laws of the United States by U.S. Deputy Surveyor W.A. Barefoot, pursuant to his contract of April 4, 1861.
>
> 2.   The returns of Barefoot's field notes reflect that he purportedly conducted his meander-line survey of the right bank of the Missouri River in the vicinity of the disputed property on January 3–5, 1862.
>
> 3.   The testimony and evidence received by the court in this cause amply establishes that Mr. Barefoot was a mendacious, licentious, and inebriate fellow, a fact which raises certain threshold questions concerning the veracity of entries in his field notes.
>
> 4.   These doubts find solace in the fact that Barefoot places the river in 1861 not closer than 15 miles to any other position the river is known to have occupied in history . . . .

Typical conclusions of law in such a case might read as follows:

> 1.   Meander lines are run in surveying fractional portions of the public lands bordering upon navigable rivers, not as boundaries of the

---

2.   Fed. R. Civ. P. 52(a).

tract, but for the purpose of defining the sinuosities of the banks of the stream, and as the means of ascertaining the quantity of land in a fractional section subject to sale, which is to be paid for by the purchaser.

2.   In preparing the official plat from the field notes, the meander-line is represented as the border-line of the stream, and shows, to a demonstration, that the water-course, and not the meander-line, as actually run on the land, is the boundary. *St. P. & P.R.R. Co. v. Schurmier*, 74 U.S. (7 Wall.) 272, 286–87 (1868).

3.   A well-established exception to this rule exists when through gross error or fraud there is no channel or body of water anywhere near the meander-line run by the surveyor. This is the situation commonly referred to as creating "omitted lands," that is, public lands erroneously or fraudulently omitted from a public lands survey. *Jeems Bayou Fishing & Hunting Club v. United States*, 260 U.S. 561 (1922), and *Producers Oil Co. v. Hanzen*, 238 U.S. 325 (1915).

In many courts, including the federal system, findings and conclusions are often stated formally, as illustrated by the examples given above. It suffices, though, if the court prepares a less rigid, narrative statement of its reasons for its decision, so long as what the court finds to be the facts and the law appears clearly from such a memorandum. In California, for example, a former statute required separately stated findings of fact and conclusions of law; California Code of Civil Procedure section 632 now requires merely a "statement of decision explaining the factual and legal basis for its decision as to each of the principal controverted issues at trial."

A statement of findings and conclusions is mandatory in all nonjury cases in the federal courts; in other jurisdictions it is either (a) discretionary with the judge or (b) mandatory when such a statement is requested by one of the parties.[3]

In state-court systems, it is frequently customary for the court to assign one of the parties, often the prevailing party, to prepare a proposed statement of findings and conclusions. The Local Civil Rules for the United States District Court for the Central District of California (Los Angeles), for example, provide in Rule 14.3:

In all cases where findings of fact and conclusions of law are required under F.R.Civ.P. 41, 52, and 65, the attorney for the prevailing party shall within five (5) court days lodge proposed findings of fact which:
(a)   Are in separately numbered paragraphs';
(b)   Are in Chronological order; and
(c)   Do not make reference to allegations contained in pleadings.
Conclusions of law shall follow the findings of fact and:
(i)   Shall be in separately numbered paragraphs and
(ii)   May include brief citations of appropriate authority.

---

3.   Compare Federal Rule 52(a) with California Code of Civil Procedure section 632.

The party who did not prepare the proposed findings of fact and conclusions of law, of course, has the opportunity to object to any portions of the proposed findings and conclusions and is frequently permitted to argue his objections orally before the court.[4]

MOTION FOR STAY OF EXECUTION

"Execution" means something wildly different in murder cases from what it means in civil cases—the cases with which we are concerned in this book. When in civil cases judgment has been entered, whether in a jury or nonjury case, the prevailing party is entitled to "execution" of the judgment. That means the carrying out of the judgment or the obtaining of the relief he was awarded, regardless of whether the losing party intends to appeal the decision. Some period of delay is normally provided for. In the federal courts, for example, no execution on a judgment may be had until the expiration of 10 days following the entry of judgment.[5]

There are certain methods available to the losing party to avoid having to comply with the judgment immediately—certainly an understandable goal if he intends to appeal. A court order which thus protects him from having to comply with the judgment is a "stay of execution," an expression which bears a grisly identity with the one employed in the criminal law.

In an action for money damages, for example, if the plaintiff has prevailed, upon entry of judgment he becomes known as the "judgment creditor," and the defendant the "judgment debtor." The defendant may wish a stay of execution on his obligation to pay the judgment, hoping that he may obtain a reversal when he appeals. In an ejectment case, the losing defendant may likewise wish to avoid having to relinquish possession of the premises to the plaintiff while he pursues a motion for new trial or an appeal. The Federal Rules of Civil Procedure provide that in the federal district courts, the trial court may stay execution of the judgment pending a motion for new trial; and they further provide that where a money judgment operates as a lien upon real property under state law, the judgment debtor in a federal case is entitled to a stay of execution under the terms of that state's law.[6] In California, whose court rules are typical of many states, the trial court has authority to stay execution of judgment, if a motion for new trial or for judgment notwithstanding the verdict is pending, until 10 days after the determination of those motions. In other situations, generally speaking, the trial court has the authority to stay execution of judgment for not longer than 30 days.

This initial authority of the trial court to stay an execution of judgment temporarily is then supplemented by rules providing for the stay of execution pending appeal. (Obviously, if the losing party does not intend to appeal, there is no purpose to be served by the staying of execution beyond the rather limited periods permitted the trial court.) To secure a stay of execution of judgment pending appeal, virtually all courts require the losing party to post an undertaking in the form of a surety bond or otherwise for the protection of the judgment creditor. The Federal Rules of Appellate Procedure provide that a stay of execution pending appeal is to be sought in the first instance in the trial court. Rule 7 of the Federal Rules of Appellate Procedure provides that the "district court may require an appellant to file a bond or provide other security in such form and amount as it finds necessary to ensure payment of

---

4. *See, e.g.,* Fed. R. Civ. P. 52(b); Cal. Civ. Proc. Code § 632.
5. Fed. R. Civ. P. 62(a).
6. Fed. R. Civ. P. 62(b) and (f).

costs on appeal in a civil case." Rule 8 governs the situation where the appellant is unable to secure a stay of execution from the district court, and likewise provides that such a stay may be conditioned upon the giving of a bond or other undertaking.

Similarly, California Code of Civil Procedure section 917.1 provides in subsection (a), "[t]he perfecting of an appeal shall not stay enforcement of the judgment or order in the trial court if the judgment or order is for money, " unless "an undertaking is given." Subsection (b) goes on to provide that the undertaking shall be on condition that if the appellant loses or dismisses his appeal, he shall pay the face amount of the judgment, together with interest that may have accrued pending the appeal and costs of the appeal that may be awarded against him. Significantly, this statute provides that the undertaking shall be for double the amount of the judgment "unless given by an admitted surety insurer in which event it shall be for one and one-half times the amount of the judgment or order."

ORDERS AWARDING LITIGATION COSTS, ATTORNEYS' FEES, ETC.

A substantial amount of post-trial maneuvering frequently centers on the question of whether and in what amount the prevailing party is entitled to be reimbursed for his litigation "costs" and, in exceptional cases, his attorneys' fees. Rule 54(d) of the Federal Rules of Civil Procedure provides that costs are usually awarded "of course" to the prevailing party in a lawsuit. "Costs," as might be evident from its use in conjunction with "attorneys' fees" at the beginning of this paragraph, is a term that does not comprise attorneys' fees. Nor does it include such matters as the attorney's travel expenses, and so on. It does include such diverse expenses of litigation as the fees charged by the court for filing the lawsuit or filing an appearance in the case, certain costs of preparing deposition transcripts, the expenses for food and lodging of the jury, and so on.[7]

Attorneys' fees are ordinarily not recoverable by the prevailing party in a lawsuit except where a statute specifically makes such fees recoverable, or where the action was upon a contract that contained a provision that should litigation be required to secure the performance of an obligation created by the instrument, the prevailing party was to be awarded attorneys' fees. California Code of Civil Procedure section 1021, for example, provides the following:

> Except as attorney's fees are specifically provided for by statute, the measure and mode of compensation of attorneys and counselors at law is left to the agreement, express or implied, of the parties; but parties to actions or proceedings are entitled to costs, as hereinafter provided.

While the federal rule provides that costs are to be awarded as a matter of course to the prevailing party in a civil suit, state-court systems provide for the awarding of costs to the prevailing party as a matter of course in some instances and as a matter of discretion with the trial judge in others. In California, costs are allowed as a matter of course to the prevailing party in actions for the recovery of real property, actions for money or damages, and actions involving the title to or possession of real property.

---

7. *See, e.g.,* Cal. Civ. Proc. Code § 1033.5.

By statute, attorneys' fees may be awarded in special cases. In California, the "private attorney general statute," Code of Civil Procedure section 1021.5, provides that a court may award attorneys' fees to the successful party in a case that has resulted in the enforcement of an important public right, provided certain conditions are present.[8]

Attorneys' fees are typically awarded to a defendant in a condemnation action in the discretion of the court. If the court finds following trial that the condemnee's pretrial offer had been reasonable and the condemnor's had not, the condemnee will receive his attorneys' fees.[9] It was mentioned above that in the federal system, costs are awarded under Federal Rule 54 as "of course" to the prevailing party. Federal Rule 71A(l) provides an exception for this rule in condemnation cases. In almost all cases, of course, the plaintiff–condemnor, being awarded the property, "prevails."

The party awarded costs is required, typically within a short period of time, to file a memorandum itemizing each of his asserted costs. The opposing party then has the privilege of filing a motion to "tax" costs, by which it is meant merely that he challenges the validity of some or all of the costs claimed by, the prevailing party. Often a hearing is held to determine the validity of the contested cost items, and when they have finally been ascertained, the precise amount awarded is then entered by the clerk of the court in the judgment, where typically a blank had been left for the purpose: "Finally, the court decrees that costs in the amount of $_____ are awarded to plaintiff."

Many federal environmental statutes provide for an award of attorneys fees in "citizens suits," such as under the Endangered Species Act and the Clean Water Act.[10] In addition, some successful plaintiffs suing the federal government may recover their lawyers' fees under the Equal Access to Justice Act.[11]

MOTIONS FOR JUDGMENT NOTWITHSTANDING THE VERDICT (NON OBSTANTE VERDICTO)

After a jury renders its verdict, a judgment *non obstante verdicto* or "notwithstanding the verdict" may be entered on motion of the losing party where it appears to the court that a directed verdict in that party's favor ought to have been ordered. By contrast, a directed verdict is in order, upon motion by the defendant at the conclusion of the plaintiff's case, or on motion of either party at the conclusion of all of the evidence, in very limited circumstances. (In federal court, a judgment notwithstanding the verdict or a directed verdict is, under Federal Rule of Civil Procedure 50, called "judgment as a matter of law.") The power of the directed verdict allows the court to take from the jury cases in which the facts are so clear that the law requires a particular result.[12] The procedure for moving the court to enter such a judgment is in most jurisdictions provided by statute.[13]

Questions regarding the constitutionality of this concept—addressing the fact that the losing party is deprived of his "right" to a jury trial—are well-settled. The right to a jury trial,

---

8. For a California Supreme Court decision declining to award attorneys' fees under this statute see *Pacific Legal Foundation v. California Coastal Com.*, 33 Cal.3d 158 (1982).

9. Cal. Civ. Proc. Code § 1250.410 (b).

10. *See* 16 U.S.C. § 1540(g)(4) and 33 U.S.C. § 1365(d).

11. 5 U.S.C. § 504.

12. Wright & Miller, Federal Practice and Procedure: Civil § 2521, at 537 (1971).

13. *See, e.g.,* Fed. R. Civ. P. 50; Cal. Civ. Proc. Code § 629.

as that right existed at common law, is the right preserved by the Seventh Amendment to the United States Constitution. It was well established at common law that although questions of fact must be decided by the jury and may not be reexamined by the court, the question whether there is sufficient evidence to raise a question of fact to be presented to the jury is a question to be decided by the court. The United States Supreme Court approved the granting of a directed verdict (i.e., taking a case from the jury) as early as 1850,[14] and the question was finally put to rest by the Supreme Court's 1943 decision in *Galloway v. United States.*[15]

THE MOTION FOR NEW TRIAL

One of the most frequently employed post-trial procedures is the motion of a losing party for a new trial. A new trial has been defined as "a re-examination of an issue of fact in the same court after a trial and decision by a jury, court or referee."[16] By definition, then, a new trial is not available where the facts of the case were not tried, as when the case has been determined upon demurrer or the default of one of the parties.[17]

The grounds for a new trial are fairly uniform among the jurisdictions of the United States. Some examples from California courts are as follows[18]:

1. *Irregularity in the Proceedings Which Prevented a Party from Having a Fair Trial.*[19] The irregularity may consist of either (a) any improperly composed jury (such as too many or too few jurors) or (b) misconduct of one of the parties or his attorney. It is ordinarily improper, for example, for a plaintiff's attorney to refer in the presence of the jury to the fact that the defendant is covered by a policy of insurance.[20] In another case, it was held to be misconduct of counsel to make allusions to the poverty of the plaintiff and the wealth of the defendant.[21] The California Supreme Court has equivocated whether passing candy and cigars to jurors is misconduct on the part of a lawyer.[22]

2. *Misconduct of the Jury.* The jury can commit misconduct, thus providing a ground for a new trial, by improperly receiving evidence (as by doing independent investigation outside the courtroom) or receiving improper communications about the trial from outsiders, as well as from the parties or the attorneys themselves.[23] Another form of jury misconduct occurs when the jury renders a "chance verdict," as by abandoning its deliberation and flipping a coin.[24]

---

14. *Parks v. Ross,* 52 U.S. (11 How.) 362 (1850).
15. 319 U.S. 372 (1943).
16. Cal. Civ. Proc. Code § 656; *Pasadena v. Superior Court,* 212 Cal. 309, 313, 298 P. 968 (1931). Fed. R. Civ. P. 59.
17. *Foley v. Foley,* 120 Cal. 33, 36 (1898).
18. The Federal Rule provides that a new trial shall be granted "for any of the reasons for which new trials have heretofore been granted in actions at law in the courts of the United States." Fed. R. Civ. P. 59 (a).
19. *See, e.g.,* Cal. Civ. Proc. Code § 657(1); *Gray v. Robinson,* 33 Cal.App.2d 177, 182, 91 P.2d 194 (1939).
20. *Gray v. Robinson, supra* note 18, 33 Cal.App.2d at 183.
21. *Weaver v. Shell Oil Co. of Cal.,* 129 Cal.App. 232, 18 P.2d 736 (1933).
22. *Whitfield v. Roth,* 10 Cal.3d 874, 519 P.2d 588 (1974).
23. *See, e.g., Wright v. Eastlick,* 125 Cal. 517, 520, 58 P. 87 (1899); Cal. Civ. Proc. Code § 657 (2).
24. *Dixon v. Pluns,* 98 Cal. 384 (1893).

3. *Newly Discovered Evidence.*   On rare occasions, newly discovered evidence, mate- rial to the case of the party applying for a new trial, which could not with reason- able diligence have been discovered and produced at the trial, provides a ground for a new trial.[25] Not surprisingly, a new trial on the ground of newly discovered evi- dence is ordinarily looked upon with "distrust and disfavor."[26]

4. *Accident or Surprise.*[27]   What is meant by "accident or surprise" is aptly illus- trated by a 1930 California decision in which an attorney was retained to represent the defendant, filed an answer, and then failed to notify his client of the trial and effectively abandoned the case.[28] It is, as might be expected, a seldom successful ground for requesting a new trial.

5. *Insufficient Evidence to Support the Verdict or Decision.*   If the verdict or decision is, in the view of the court, against the *weight* of the evidence, a new trial may be ordered.[29] This criterion gives the trial judge hearing a motion for new trial more latitude than an appellate court to set aside a verdict for inadequate evidence. An appellate court, as explained below, is required to confine itself to a consideration of whether there was any "substantial" evidence to support the decision; it may not consider whether the verdict was merely against the "weight" of the evidence.

6. *Excessive or Inadequate Damages.*   This ground is in effect a corollary to the ground of insufficient evidence. If the court is convinced, upon a motion for trial, that the damages awarded were excessive, for example, it will commonly order a new trial unless the plaintiff stipulates to a lesser amount of damages. Similarly, if the plaintiff has moved for a new trial on the ground that the awarded damages are inad- equate, and the trial court is convinced of the merit motion, it will enter an order granting a new trial unless the defendant agrees to have the amount of damages in- creased. The former situation is referred to as "remittitur" and the latter as "additur."

MOTIONS FOR RELIEF FROM JUDGMENTS BASED ON EXCUSABLE NEGLECT, ETC.

Parties and their counsel occasionally find that orders and even judgments have been entered against them without their knowledge. Under very limited circumstances, relief may be had in these situations. Rule 60(b) of the Federal Rules of Civil Procedure provides, for example:

> On motion and upon such terms as are just, the court may relieve a party or a party's legal representative from a final judgment, order, or proceeding for the following reasons: (1) mistake, inadvertence, surprise, or excusable neglect. . . . The motion shall be made within a reasonable time, and . . . not more than one year after the judgment, order, or proceeding was entered or taken.

---

25. *See, e.g.,* Cal. Civ. Proc. Code § 657(4).
26. *See, e.g., Estate of Emerson,* 170 Cal. 81, 82, 148 P. 523 (1915).
27. Cal. Civ. Proc. Code § 657(3).
28. *People's F. & T. Co. v. Phoenix Assur. Co.,* 104 Cal.App. 334, 285 P. 857 (1930).
29. *See, e.g.,* Cal. Civ. Proc. Code § 657(6).

The same language is found in the statutes of many states, including California.[30] As ought to be expected, it takes a compelling case to secure relief under such provisions.

B.   APPEAL

An appeal from a decision of the trial court is initiated in virtually all jurisdictions by the filing of a "notice of appeal" within the time permitted by the applicable rules. In the federal-court system, for example, the notice of appeal must be filed with the district court (i.e., the trial court) within 30 days after the date of the entry of judgment in the ordinary case; in the case of the federal government as appealing party, the notice of appeal need be filed within 60 days after the date of the entry of judgment.[31] In state court systems, it is frequently the rule that the notice of appeal must be filed within 60 days of either the date of the entry of judgment or the date that notice of the entry of judgment was mailed.[32]

When the notice of appeal has been filed, the next step in the appellate process is the preparation of the "record on appeal." Typically, this record consists of the *clerk's* record (essentially the pleadings and papers filed with the court) and the *reporter's* transcript of proceedings. This transcript includes a verbatim text of the testimony at trial, including the comments of the judge and the lawyers, exhibits introduced at the trial, and occasionally transcripts of pretrial hearings.

When the record has been lodged with the appellate court, the appellant (the loser of the decision below) then files his opening brief, and the appellee (the prevailing party in the court below) files his responding brief; typically the appellant has the opportunity to close the briefing with a reply brief.

The matter is then set down for oral argument, following which a decision is rendered with greater or less speed.

There is a prevalent but mistaken belief that an appeal represents in effect a chance to try the case again. Ambrose Bierce apparently thought so. In fact, though, when a case is briefed and argued on appeal, very little in the way of new evidence may be introduced (in general, only matters that may be judicially noticed), and with one exception the only issues are questions of law: Did the trial court err in applying the law to the facts? (The exception, the lack of sufficient evidence to support the decision in the trial court, is treated below.) Moreover, even if such an error is found, it must be shown to have been "prejudicial"; that is, there must be a substantial likelihood that the result would have been different had the error not occurred. Some discussion of the nature of the judicial system in the United States is helpful to an understanding of what occurs on an appeal from a decision at trial.

SOME PERTINENT CHARACTERISTICS OF OUR SYSTEM OF GOVERNMENT
Its Federalist Nature

One distinguishing feature of the system of government and hence of the judicial systems created in the United States Constitution is its "federal" nature. Federalism is the

---

30.   Cal. Civ. Proc. Code § 473.

31.   Fed. R. App. P. 4.

32.   *See, e.g.,* Cal. R. Ct. 2 (a).

division of political power between a central government, with certain authority over the entire territory of the nation, and states or local governments with geographically limited authority. This dual sovereignty between the states and the federal government accounts for the diversity of legal principles in the areas of land title and boundaries, among other areas.

A National Government of "Enumerated Powers"

Federalism thus involves a division of powers between the national and state governments; to accomplish this, the framers of the Constitution devised a plan by which the powers of the National Government were enumerated in somewhat precise fashion, leaving it to be inferred that all remaining powers were reserved to the states. This fundamental principle dividing powers between the federal government and the states was expressly added to the Constitution in the Tenth Amendment. It is frequently said that the federal government is one of but enumerated powers.

The power of Congress to control navigation illustrates this point. The Constitution says nothing of navigation in so many words, but Article I, Section 8, clause 3 grants to Congress the power "To regulate commerce with foreign nations, and among the several States, and with the Indian tribes." Clause 18 of the same Article and Section gives Congress the power to legislate everything "necessary and proper" to carry out its expressly designated powers. These clauses were construed by the United States Supreme Court in the 1824 case of *Gibbons v. Odgen*[33] to apply to the regulation and protection of navigation upon navigable interstate rivers as a "necessary and proper" incident of the power to regulate commerce.

An Exception: The Foreign-Relations Power

A salient exception to the principle that the national government is one of enumerated powers only is found in the field of foreign relations. Nowhere in the Constitution is the power to conduct foreign relations expressly committed to the federal government. (It is, though, off limits to the states.[34]) The United States Supreme Court has held nevertheless that in the conduct of its foreign affairs the United States as a sovereign nation holds all of the powers that sovereignty necessarily implies, and that these powers are hence not limited to those that are expressly delegated to it by the Constitution. The court has distinguished the powers of the federal government respecting domestic affairs from those respecting foreign affairs:

> In that field [internal affairs], the primary purpose of the Constitution was to carve from the general mass of legislative powers *then possessed by the states* such portions as it was thought desirable to vest in the Federal government, leaving those not included in the enumeration still in the states. . . . That this doctrine applies only to powers which the states had, is self-evident. And since the states severally never possessed international powers, such powers could not have been carved from the mass of state powers but obviously were transmitted to the United States from some other source.[35]

---

33.   22 U.S. (9 Wheat.) 1 (1824).
34.   *See* U.S. Const., Art. I, § 10, cl. 1.
35.   *United Staes v. Curtiss-Wright Export Corp.*, 299 U.S. 304, 316 (1936).

There are thus powers that are exclusively the national government's, powers that are exclusively the states' (pursuant to the Tenth Amendment to the Constitution), and powers that may be exercised concurrently. Powers such as the foreign relations power must by their very nature be exercised solely by the national government. Other powers committed to the federal government, such as the power to regulate interstate commerce, may under certain conditions be exercised by the states if the national government has not legislated in the area, for example through its enumerated power contained in the commerce clause. Among the subjects that, in the absence of federal legislation, are within state jurisdiction are ferry franchises, bridge construction, and pilotage and harbor regulations.[36] The question of the power to make and enforce rules of property law, a most complex subject, is dealt with in the preceding chapter.

The Separation of Powers

Another fundamental principle of the American system of government, found at both state and federal levels, is that legislative, executive, and judicial powers are vested in separate branches of the government. This tripartite version of government operates on the premise that different bodies, for the most part, should make the laws, enforce them, and adjudicate disputes based upon them, whether criminal or civil. The principle of the separation of powers finds its origin in the United States Constitution which provides that "All legislative powers herein granted shall be vested in a Congress of the United States," that "The executive power shall be vested in a President of the United States of America," and that "The judicial power of the United States shall be vested in one Supreme Court and in such inferior courts as the Congress may from time to time ordain and establish."[37]

With respect to the form of state governments, the federal Constitution requires only that the states guarantee a "republican form of government," that is, a government of representatives.[38]

There is not an absolute separation of powers, of course, because of the system of "checks and balances" found in the Constitution; each branch functions to exert a certain restraint upon the others. The veto power of the President, the power of Congress to override a veto, and the power of the Supreme Court to hold unconstitutional acts of Congress and acts of the President are examples.

The separation-of-powers principle had become almost a hackneyed, high-school civics-class subject until Watergate. In the wake of Watergate, Congress enacted the Ethics in Government Act. That Act provides for the appointment, by judges, of an Independent Counsel, a "special prosecutor," for cases in which an investigation of the Executive Branch must be made, but in which the appearance, at least, of a conflict of interest is very great. (Ordinarily such investigations are made by the Department of Justice, which is headed by the Attorney General, who is appointed by the President.) In the 1980s the Act was challenged by Ted Olsen, Oliver North, and others as a violation of the principle of the separation of powers. The argument, simply put, was that the execution of the laws is committed by our Constitution to the President—the Executive Branch of our national government—and it is thus unconstitutional to invest that power in the judicial branch. The Supreme Court rejected these

---

36. *See* cases collected in *Covington & Cincinnati Bridge Co. v. Kentucky,* 154 U.S. 204 (1894).
37. U.S. Const., art. I, II, and III, respectively.
38. U.S. Const., art. IV, § 4.

challenges in a case brought against Independent Counsel Alexia Morrison. Ms. Morrison and the Ethics in Government Act were ably represented before the Court by my partner and beloved friend Louis Claiborne [*Morrison v. Olson*, 487 U.S. 654 (1988)]. The decision was, for Ms. Morrison and our office, most gratifying, of course. It was perhaps the most important constitutional-law decision of the Supreme Court's term, and it was nearly unanimous; only Justice Antonin Scalia dissented. Truth be told, Scalia's passionate dissent made the case for the unconstitutionality of the Act far better than had the lawyers for Ted Olsen and Oliver North. What is more, he raised vividly the specter of some future Independent Counsel run amok. The ultimate outcome of the Whitewater Independent Counsel's endeavors may demonstrate whether Scalia had the better of the point, to the Nation's sorrow.

More recently the citizens-suit provision of various federal environmental laws have been challenged on similar grounds. The Clean Water Act is a prime example. Most water-discharge permits issued to city sewage-treatment facilities contain effluent limitations that cannot be met. (No chlorine whatever may be discharged, for example.) The responsible agencies won't sue to enforce these impossible, "technology-driven" standards, but citizens can; and they can seek penalties of $25,000 per day, and handsome fees for themselves to boot. The separation-of-powers argument is that the President (the Executive), and not you or I, is the one person constitutionally empowered to enforce the laws.

THE NATURE OF THE JUDICIAL SYSTEM IN THE UNITED STATES

Common Characteristics of the State and Federal Judiciaries

The legislature, then, makes law, and it is the role of the courts to apply the law to specific situations. The structure of the judicial system mirrors the dual government system. Just as there are two levels of legislature, federal and state, there are two separate court systems existing side by side, the federal judiciary and the state judiciaries. This dual sovereignty in the legislative and judicial fields accounts for the lack of uniformity in the laws relating to land title and boundaries, riparian rights, and related matters.[39] This situation makes it impossible to state what "the rule" is in the United States with respect to a property or boundary matter, and makes it necessary at the least to distinguish between what may be the rule in federal courts and what rule a state court may apply.

A second characteristic of the judicial structures of both the federal government and the state governments is the hierarchical scheme. At the base of the judicial apex are the trial courts, or courts of general authority, to hear and decide virtually all cases in the first instance. These courts hear the evidence, ascertain the facts, and apply the law. Above these trial courts are intermediate appellate courts, whose purpose is to secure a uniform interpretation of the law. The appeal of cases in which the meaning of law is involved provides the mechanism for handing down authoritative rulings on doubtful legal issues, which then must be followed by the lower courts. Appellate courts function without juries, and no new

---

39. "The settled decisions of the courts of a state and its laws . . . determine the title to the beds of navigable streams and the extent of rights of riparian owners . . . in that state. *Iowa v. Carr*, 191 F. 257, 261 (1911). If [the states] choose to resign to the riparian proprietor rights which properly belong to them in their sovereign capacity, it is not for others to raise objections. *Barney v. Keokuk*, 94 U.S. (4 Otto) 324, 338 (1877).

evidence, with the exception of matters that may be judicially noticed, is presented. Both the federal judiciary and most states have a third tier of courts consisting of a supreme court. Alaska is one state that only has courts of general jurisdiction (trial courts) and a state supreme court.

The Federal Judiciary

The federal judiciary is created by Article III, Section 1 of the United States Constitution, which provides, the following: "The judicial power of the United States shall be vested in one Supreme Court and in such inferior courts as the Congress may from time to time ordain and establish." Under the terms of the Constitution, the federal courts may adjudicate only cases arising under the Constitution, laws and treaties of the federal government (so-called "federal question" cases), admiralty and maritime cases, cases affecting ambassadors, ministers and consuls, controversies between two or more states, cases in which the United States is a party controversies between citizens of different states (so-called "diversity" cases), controversies between a state, or the citizens thereof, and foreign states, citizens, or subjects, disputes between a state and citizens of another state, and controversies between citizens of the same state claiming lands under grants of different states. All other cases are the domain of the state courts.

The Constitution, however, does not mandate that all of these classes of cases be heard solely by the federal courts. Congress has the authority to specify cases that may be tried either in federal or state courts and can permit the state courts to hear certain kinds of cases on a concurrent or even exclusive basis. The Eleventh Amendment to the Constitution provides that "The judicial power of the United States shall not be construed to extend to any suit in law or equity, commenced or prosecuted against one of the United States, by citizens of another State, or by citizens or subjects of any foreign state." This amendment has been construed as well to prevent a state from being sued without its consent by its *own* citizens[40] or by a foreign state.[41] That is not to say, however, that one cannot sue state *officers,* who have violated the federal civil rights laws, for example.

The Supreme Court has held that the federal courts, except where the case is controlled by the Constitution or acts of Congress, must apply the law of the state in which it sits. (There are not infrequently questions whether another state's law must be applied.) Decisions of the highest state court are, of course, the best authority as to state law.[42]

The trial courts or courts of general jurisdiction of the federal judiciary are the district courts, first established by the Judiciary Act of 1789 (1 Stat. 73). Each state has at least one federal district court, some states have more than one, but no judicial district crosses a state boundary. The District of Columbia also has a district court of its own. These are the only federal courts where juries are used—unless the Supreme Court chooses to use one in its original jurisdiction. They have the "original" jurisdiction of civil cases (that is, authority to hear the cases first) at common law, in equity, in admirality, and in enforcement of acts of Congress, and of all prosecutions for crimes under federal law, but have no jurisdiction to hear appeals.

The second tier of the federal judiciary is the circuit courts of appeals. There is one such court in each of the 11 judicial "circuits" into which the country is divided for this purpose.

---

40.   *Hans v. Louisiana,* 134 U.S. 1 (1890).

41.   *Monaco v. Mississippi,* 292 U.S. 313 (1934).

42.   *Erie R. Co. v. Tompkins,* 304 U.S. 64, 80 (1938).

The District of Columbia also has its own appellate court known as the United States Court of Appeals for the District of Columbia. The jurisdiction of these courts is exclusively appellate; they have no original jurisdiction to hear cases. They hear civil and criminal appeals from the district and other federal courts (such as bankruptcy courts) that sit within each circuit. Created by Congress in 1891,[43] they serve to alleviate the workload of the Supreme Court, as well as to filter cases so that only the most significant ones seek review by the highest tribunal. The courts of appeals cannot review the decisions of state courts, these decisions, if they are reviewed at all in the federal judiciary, are reviewed by the Supreme Court. In cases in which a court of appeals has held a *state* statute invalid on the ground it is in conflict with the Constitution, a law, or a treaty of the United States, an appeal may be taken to the Supreme Court. (While such an appeal is by law a matter of "right," exercising one's right to appeal to the Supreme Court in such cases does not ensure a full hearing by that court. In cases that the court feels are not sufficiently important, considering its workload, to hear oral argument and render a full decision, the appeal is summarily disposed of by a short one-paragraph order affirming the decision of the court of appeals, or reversing it, usually with a brief citation to the authority for the reversal.) In all other cases, the decisions of the courts of appeals are final except as they may be reviewed in the discretion of the Supreme Court, not by "appeal" but by issuance of a "writ of certiorari."

At the apex of the federal judicial pyramid is the Supreme Court, the only federal court specifically provided for in the Constitution. The Supreme Court's authority to hear cases that are decided by the federal courts of appeals was discussed briefly in the preceding paragraph. In addition, the Supreme Court may accept cases for review directly from the state courts in certain situations. Such situations include cases in which the validity of a state or federal statute under the United States Constitution is called into question. The Supreme Court's appellate jurisdiction—that is, its authority to hear cases previously decided by federal or state courts—may be diminished by Congress, but its authority under the Constitution to hear cases in the first instance, its "original jurisdiction," cannot be. (Congress can, however, grant "concurrent" jurisdiction to lower courts, as it has in the case of suits between the United States and a state. In such instances the Supreme Court may, but need not, accept the case in the first instance.[44]) Original jurisdiction cases, as mentioned above, comprise for example disputes affecting ambassadors, public ministers, and consuls, or cases in which a state is a party.

In cases where the Supreme Court exercises its original jurisdiction, it rarely functions precisely as a federal district court conducting a trial. Ordinarily, the Supreme Court passes on questions of law only, through the typical course of briefing and oral argument. When the ascertainment of facts, (i.e., a trial) is necessary, these cases are ordinarily referred to a special master.[45] Such a special master has the authority to receive evidence and hear testimony and to report back his recommendations to the full court, but he has no power to decide the

---

43. 26 Stat. 826, as amended by 45 Stat. 1346 (1929).

44. 28 U.S.C. § 1251.

45. There have been no known jury trials in the Supreme Court since 1797. *See* "The Supreme Court—Its Homes Past and Present," 27 A.B.A.J., 283, 286 n.3 (1941). Section 1872 of Title 28, U.S.C., provides, "In all original actions at law in the Supreme Court against citizens of the United States, issues of fact shall be tried by a jury."

controversy, nor to rule on major issues brought by motion, and so on.[46] This procedure has been employed by the Supreme Court in a number of interstate boundary cases, water disputes, and the litigation concerning the Submerged Lands Act. For example, after the decision in *United States v. California,* 332 U.S. 19 (1947), in which the Supreme Court held that the federal government was possessed of "paramount rights" in the submerged lands lying seaward of the ordinary low-water mark on the coast of California, it named a special master to preside over hearings on the question of the precise location of the seaward boundaries of California's inland waters. The Supreme Court has employed Special Masters in many of the other submerged-lands cases.[47]

With the exception of cases in its original jurisdiction, the Supreme Court does not hear a case ordinarily until at least one and usually two lower courts have heard and decided it. Moreover, as with other appellate courts, the Supreme Court ordinarily acts only upon the facts disclosed by the "record" made in the court below (again with the exception of matters that may be judicially noticed).

With very few exceptions, as a practical matter, review in the Supreme Court is not a matter of right, but one of discretion of the court. Granting of review depends in the main upon the importance of the case to the nation as a whole, and not upon any perceived injustice to an individual litigant.

Cases reach the United States Supreme Court by five routes. The first is from state courts, when the validity of a state or federal statute under the federal Constitution is called into question. The second is from the United States District Courts in cases in which a federal statute has been ruled unconstitutional and the federal government is a party. Under such circumstances, the United States may appeal directly to the Supreme Court. The third avenue is from the Courts of Appeals where the decisions of different circuits are conflicting on the same issue, and cases in which the Supreme Court feels the Court of Appeals has misconstrued or misapplied an earlier Supreme Court decision. The final avenue is from the Court of Claims, the Customs Court, and the Court of Customs and Patent Appeals. The fifth is the Court's original jurisdiction.

One characteristic distinguishing the United States Supreme Court from the highest tribunals of most other nations is its power to hold unconstitutional an act of the President, of the Congress, or of a state. In the words of former Justice Robert Jackson of the United States Supreme Court, "that power is not expressly granted in the Constitution, but rests on logical implication. It is an incident of jurisdiction to determine what is the law governing a particular case or controversy. In the hierarchy of legal values, if the higher law of the Constitution prohibits what the lower law of the legislature attempts, the latter is a nullity."[48] For reasons of constitutional doctrine as well as practicality, however, there is a considerable reluctance on the part of the Supreme Court to hold acts of Congress or of the President unconstitutional. Aside from the deference that is to be accorded the coequal branches of government, there is the very pragmatic consideration that the Supreme Court is a relatively frail institution, having no independent power to finance itself (that is the prerogative of Congress) and rela-

---

46.   R. Stern & E. Gressman, Supreme Court Practice, at 488 (7th ed., 1993).

47.   *See* M. Reed, G.T. Koester, & J. Briscoe, Reports of the Special Masters in the Submerged Lands Cases (1991).

48.   Jackson, *The Supreme Court in the American System of Goverment* at 22 (1955).

tively little in the way of personnel to secure by force its orders and decrees. Consequently, if a case can be resolved without the Court's reaching the question of the constitutionality of an act of Congress or the President, the Court will inevitably do so.

With particular respect to boundary cases, it might be noted that Congress by the Constitution was made an overseer of interstate compacts, many of which have been entered into for the purpose of resolving boundary disputes (Art. I, § 10), and the Supreme Court was made the arbiter of disputes between states (Art. III, § 2). In this field, both when there are interstate compacts and when there are not, the court has adjudicated many disputes concerning interstate river boundaries.[49]

The number of cases filed each year in the United States Supreme Court has grown increasingly in the past 50 years. The number of filings increased from 1181 in the 1950 term to nearly 6000 in the 1991 term.[50] During the 1976 term, of the 3889 cases acted upon by the Court, nearly 4.4 percent were disposed of by written opinion following oral argument, 5 percent were decided summarily, and 90.5 percent were decided, dismissed or withdrawn.[51] By 1991, oral argument was being held in just slightly more than 2 percent of the cases filed.[52] The success of the United States Solicitor General, which represents the federal government in nearly all actions involving the United States in the Supreme Court, is considerably greater than the average litigant. The Solicitor General's success rate has in recent years been about 75 percent. His success in this respect is due to both (a) the fact that cases involving the federal government are apt to be of greater general public importance and (b) the strictness with which his office screens cases that the government has lost before deciding to petition for *certiorari*.[53]

Finally, remarks made by Chief Justice Vinson in 1949 respecting the role of the Supreme Court are instructive:

> During the past term of the Court [fifty years ago], only about 15% of the petitions for certiorari were granted, and this figure itself is considerably higher than the average in recent years. While a great many of the 85% that were denied were far from frivolous, far too many reveal a serious misconception on the part of counsel concerning the role of the Supreme Court in our federal system. I should like, therefore, to turn to that subject very briefly.
>
> The Supreme Court is not and never has been, primarily concerned with the correction of errors in lower court decisions. In almost all cases within the Court's appellate jurisdiction, the petitioner has already received one appellate review of his case. The debates in the

---

49.   *See, e.g., Oklahoma v. Texas,* 260 U.S. 606, 631 (1923); *Georgia v. South Carolina,* 257 U.S. 516 (1922); *Iowa v. Illinios,* 147 U.S. 1 (1893); *New Jersey v. Delaware,* 291 U.S. 361 (1934); *Washington v. Oregon,* 211 U.S. 127 (1908); *Maryland v. West Virginia,* 217 U.S. 1 (1910); *Indiana v. Kentucky,* 136 U.S. 479 (1890); *Alabama v. Georgia,* 64 U.S. (23 How.) 505 (1860); *Arkansas v. Tennessee,* 269 U.S. 152 (1925).

50.   R. Stern, E. Gressman, S. Shapiro, and K. Geller, Supreme Court Practice, at 33 (7th ed., 1993).

51.   R. Stern and E. Gressman, Supreme Court Practice, at 259 (5th ed., 1978).

52.   R. Stern, E. Gressman, S. Shapiro, and K. Geller, Supreme Court Practice, at 33 (7th ed., 1993).

53.   R. Stern and E. Gressman, Supreme Court Practice at 261 (5th ed., 1978).

Constitutional Convention make clear that the purpose of the establishment of one supreme national tribunal was, in the words of John Rutledge of South Carolina, "to secure the national rights & uniformity of Judgments." The function of the Supreme Court is, therefore, to resolve conflicts of opinion on federal questions that have arisen among lower courts, to pass upon questions of wide import under the Constitution, laws, and treaties of the United States, and to exercise supervisory power over lower federal courts. If we took every case in which an interesting legal question is raised, or our *prima facie* impression is that the decision below is erroneous, we could not fulfill the Constitutional and statutory responsibilities placed upon the Court. To remain effective, the Supreme Court must continue to decide only those cases which present questions whose resolution will have immediate importance far beyond the particular facts and parties involved. Those of you whose petitions for certiorari are granted by the Supreme Court will know, therefore, that you are, in a sense, prosecuting or defending class actions; that you represent not only your clients, but tremendously important principles, upon which are based the plans, hopes and aspirations of a great many people throughout the country.[54]

State Court Systems

The system of state courts found throughout the United States follows in nearly all cases the structure of the federal judiciary, with courts of general jurisdiction (trial courts), intermediate appellate courts, and final appellate courts variously denominated the "California Supreme Court," the "New York Court of Appeals," and so on.

These state courts essentially hear cases arising under laws enacted by the state legislature and under the constitutions of the states. But this is a gross overgeneralization, and it is a frequent subject of controversy whether a case may or must be heard in a state court, or may or must be heard in a federal district court. For one thing, a given case may entail matters of both federal and state law. For another, Congress has by law left a significant area of federal jurisdiction provided for in the Constitution to the domain of state courts, either concurrently with the federal courts or exclusively. Congress has provided, for example, in 28 U.S.C. section 1332, that the jurisdiction of the federal district courts in disputes between citizens of different states is limited to those cases in which the "amount in controversy" exceeds $75,000. (In 1988 the amount was raised from $10,000 to $50,000. In 1996 it was raised again, to the present amount.) In a money-damages case, then, between a plaintiff who is a resident of Missouri and a defendant who is a California resident, the case may be brought in federal court only if plaintiff is suing for more than $75,000. If the "amount in controversy" is less, the suit may be brought in a California court, a Missouri court, or perhaps even the courts of other states, depending on the circumstances of the underlying transaction giving rise to the suit. This question of jurisdiction is much too vast for more discussion here than this.

---

54. Address of Chief Justice Vinson before the Amer. Bar Assn., Sept. 7, 1949, set forth in 69 S.Ct. v., vi (1948); *see also Dick v. New York Life Ins. Co.,* 359 U.S. 437, 452–454 (1959) (Frankfurther, J. dissenting).

Cases of title to or injury to land are ordinarily brought in the courts of the state where the land is located. Environmental-damage cases, as a general rule, may be brought in the state where the damage occurred or where the defendant resides.

The appellate process in state courts typically follows the pattern of the federal judiciary. Where, as in most states, a three-tiered system exists, the losing party has a right of appeal to the intermediate appellate court. Review by the highest court of the state is then ordinarily a matter of discretion with that court; given its generally heavy workload, it will ordinarily accept for review only those cases that present questions of broad application. The relative rights of the immediate parties to the case are not of paramount concern. Where, as in Alaska, there are no intermediate appellate courts, review by the highest state court is usually considered a matter of right.

It is useful to consider some of the general principles that apply to the review of cases by all courts exercising appellate jurisdiction.

SOME GENERAL PRINCIPLES ON APPEAL

As has been mentioned, ordinarily an appellate court reviews the decision of a lower court solely to determine whether errors of law have been committed. In a relatively straightforward decision, for example, the California Court of Appeal in 1971[55] reviewed on appeal a decision of the trial court in a boundary dispute. The trial court had approved the proportionate method of locating the common boundary between the lands owned by the disputing parties. The Court of Appeal reversed the trial court's decision, holding that the application of the proportionate method was improper under the circumstances.

> [A]n examination of the record nevertheless discloses abundant evidence that the quarter corner between sections 6 and 7, as originally set by Woods [the United States Deputy Surveyor], could be located on the ground by methods other than proportionate measurement.[56]

The court went on to recite the applicable law.

> It is settled law that the proportionate measurement method may be used only as a last resort when the original corner is "lost" and cannot be relocated on the ground. A survey of public lands does not merely ascertain boundaries; it creates them. A government township lies just where the government surveyor lines it out on the face of the earth. "The line as surveyed and described in the field-notes is the description by which the government sells its land. If its description makes one section contain three hundred twenty acres and another nine hundred sixty acres, the parties must take according to the calls of their patents." [Citations omitted.][57]

---

55. *State of California v. Thompson,* 22 Cal.App.3d 368, 99 Cal.Rptr. 594 (1971).
56. 22 Cal.App.3d at 378.
57. *Id.* at 377.

To take more exotic boundary cases as examples, the United States Supreme Court held in *Borax Consolidated v. Los Angeles,* 296 U.S. 10 (1935), that the federal Court of Appeals for the Ninth Circuit had correctly applied federal and not state law to determine the boundaries of land conveyed by

> a federal patent which, according to the plat, purported to convey land bordering on the Pacific Ocean . . . . The question as to the extent of this federal grant, that is, as to the limit of the land conveyed, or the boundary between the upland and the tideland, is necessarily a federal question. It is a question which concerns the validity and effect of an act done by the United States; it involves the ascertainment of the essential basis of a right asserted under federal law."[58]

In 1973, in the case of *Bonelli Cattle Co. v. Arizona,*[59] the United States Supreme Court reversed a decision of the Supreme Court of Arizona, which had applied its state law to determine the effect on riparian boundaries of the recession of the waters of the Colorado River following a rechannelization project.

> The present case . . . does not involve a question of the disposition of lands, the title to which is vested in the State as a matter of settled federal law. The very question to be decided is the nature and extent of the title [of Arizona] to the bed of a navigable stream held by the State under the equal-footing doctrine and the Submerged Lands Act. In this case, the question of title as between the State and a private landowner necessarily depends on a construction of a "right asserted under federal law."[60]

A mere four years later, the Supreme Court overruled its decision in *Bonelli,* holding that such disputes should be governed by the laws of the state in which the property is located, and not by federal law, the equal footing Doctrine and the Submerged Lands Act notwithstanding.[61] In this case, the Supreme Court reversed the decision of the lower court, the Supreme Court of Oregon, which had confidently relied upon the recent *Bonelli* decision as guidance for its choice of the applicable law.

The decisions of appellate courts, and not trial courts, are those that operate as legal "precedent" under the doctrine of *stare decisis* (literally, "to stand decided").[62] Obviously, then, the entire body of appellate boundary cases could be cited for examples of the kinds of legal questions treated by the appellate courts on appeal.

---

58.   296 U.S. at 22.
59.   *Bonelli Cattle Co. v. Arizona,* 414 U.S. 313 (1973).
60.   *Id.* at 320–321.
61.   *State Land Bd. v. Corvallis Sand & Gravel Co.,* 429 U.S. 363 (1977).
62.   Under this principle, lower courts are constrained to follow the decisions of higher courts. A trial court, for example, must follow the decisions of all appellate courts within its system, and intermediate appellate courts must follow the decisions of the system's highest court.

Ordinarily, appellate courts engage in no attempted redetermination of the facts, accepting the facts as determined by the trial court. An appellant occasionally succeeds, however, in convincing an appellate court that the trial court's decision was based on "insubstantial evidence." The burden on an appellant advancing such an argument is quite onerous, and the reported cases are few in which a decision of the trial court is reversed on this ground.[63] This principle was explained in a decision of the California Supreme Court:

> There must be more than a conflict of mere words to constitute a conflict of evidence. The contrary evidence must be of a substantial character, such as reasonably supports the judgment as applied to the peculiar facts of the case. The rule announced in *Morton v. Mooney, et al.,* 97 Mont. 1 [33 Pac. (2d) 262], correctly states the rule which has been approved by this court in a number of our decisions. It is thus stated: "While the jurors [here the trial court was the trier of fact] are the sole judges of the facts, the question as to whether or not there is substantial evidence in support of a plaintiff's case is always a question of law for the court [citation omitted], and in determining this question 'the credulity of courts is not to be deemed commensurate with the facility and vehemence with which a witness swears. "It is a wild conceit that any court of justice is bound by mere swearing. It is swearing creditably that is to conclude the judgment." ' " [64]

Finally, as has been mentioned, ordinarily no new evidence may be presented to the appellate court, which confines its review to the facts presented when the case was tried. A seldom-applied exception to this principle provides that an appellate court may take judicial notice of relevant matters for the first time on appeal. In the *White* case cited above, the California Court of Appeal did so, with crucial effect:

> But for some reason, or perhaps no reason, the remaining contiguous survey and patent (Swamp and Overflowed Lands Survey No. 15, issued in 1883) were not offered in evidence at the trial. After affording each "party reasonable opportunity to meet such information" [citation omitted], we have taken judicial notice of this patent, an act of the chief executive officer of the state, the Governor.[65]
>
> The survey plat and patent, having been judicially noticed, proved decisive in the Court's decision, which reversed the trial court.

---

63. One of the few such cases in the boundary field is *White v. State of California,* 21 Cal.App.3d 738, 99 Cal.Rptr. 58 (1971).

64. *Herbert v. Lankershim,* 9 Cal.2d 409, 471–472, 71 P.2d 220 (1937).

65. 21 Cal.App.3d at 762.

Appendix 1

TO THE WITNESS WHO IS ABOUT
TO BE DEPOSED

The purpose of this memorandum is to provide you written background information and suggestions about depositions. A deposition is your oral testimony taken under oath by a court reporter in response to questions by the opposing attorney and, in some cases, by your attorney. The testimony is typed after the deposition is concluded and is available for use at trial by either side in certain instances permitted by the rules of civil procedure or evidence, applicable to federal or state court cases as the case may be. Neither judge nor jury is present at the deposition; only the lawyers, the witness, and a representative of each party are entitled to attend.

In effect, little difference exists between testimony at deposition and testimony at trail, except that depositions are not public, and no judge is present.

The opposing attorneys are taking your deposition for several reasons:

1. They want to determine what facts you know regarding the issues in the litigation. They are interested in knowing *now* the content of what may be your trial testimony later.

2. They want to commit you to specific testimony so that you feel you must testify at the trial in a manner consistent with your deposition testimony (or else you are in the embarrassing position of contradicting your deposition testimony). This eliminates for them some of the element of surprise.

3. They may hope to entrap you into any misstatement so that they can show at trial that you are not a truthful person whose testimony should be believed by the judge or jury.

4. They want to "look you over" under the pressure of testimony, observe your manner of answering questions, and form an impression concerning the probable effect your testimony will have on an impartial listener. Perhaps this last consideration is the most important. Lawyers are discouraged when the opposing witnesses are confident, informed, solid, and apparently unflappable. (Favorable settlements are often encouraged by the opposing attorney's belief that the witnesses against him are credible and persuasive.)

5. If you are testifying as an expert, they wish to ascertain your opinion on those matters upon which you will testify as well as the basis for your opinion. They will seek to develop all facts, documents, statements, and so on, that you have

179

considered. They will attempt to commit you to a position that can be discredited at trial by unassumable facts, documents, or the testimony of other experts. Also, they may seek to prove their case through your testimony. Or, they may attempt to create a conflict between your testimony and that of other experts testifying for the client you have been retained by.

The Federal and California Rules of Civil Procedure and Evidence Code provisions permit various uses of deposition transcripts by parties. In certain circumstances, deposition testimony may be treated as trial testimony. One of the most frequent uses is referred to as "impeachment." For instance, if you testify at your deposition that you first met Mr. XYZ in 1973 and later testify at the trial that you first met XYZ in 1975, the other lawyer is entitled to ask you:

1. If you recall his taking your deposition,
2. If you recall his asking you a particular question and your giving a particular answer, and
3. Whether you told the truth when you gave the first answer or are telling the truth in giving the second answer.

Because attorneys are permitted to utilize depositions for impeachment, the importance of precise and unemotional testimony is paramount. A good "rule of thumb" is to testify as if a judge and jury were sitting in the deposition room with you, observing your behavior, listening to the questions and answers, and deciding the case on the basis of whom they believed (and whom they liked more). A helpful perspective is to assume that the judge and jury are of diverse race, religion, and status in life. Assume that some would naturally like you and some would naturally dislike you. Attempt to say things that enforce their positive impressions and rebut their negative impressions.

The following is a series of general rules that will be helpful. Please review them carefully; they can be of great assistance, even if you have testified many times before.

1. *Remember that you have no purpose to serve other than to give the facts (or if you are an expert, opinions) as you know them.* You must give the facts if you know them. A deposition is not a trial and you are not obligated to persuade the other attorney of the merits of our position or to state our case in full. Simply stated, we want you to answer their questions and leave promptly.

2. *Never state facts (or opinions if you are an expert) that are beyond your knowledge.* Frequently, a witness is asked a question and, in spite of the fact that the witness feels he should know the answer, he does not. The witness is tempted to guess at or estimate the answer to avoid admitting lack of knowledge. This is a serious mistake. If you do not know an answer to a question, even though you think you may appear uninformed or evasive (to the other lawyer only) by stating that you do not know, you should nevertheless admit your lack of knowledge. "I guess" is generally the wrong answer and one from which your opponents can show that you either do not know what you are talking about or that you are deliberately misstating. Often, the opposing attorney knows the answer to his question but asks nonetheless, hoping you feel compelled to guess. Do not guess!

3. *Never attempt to explain or justify your answer.* If a proper question is asked, you may be required to elaborate on your answer. However, you are not required,

in argumentative fashion, to justify the rectitude or consistency of your answer. You are not to apologize for or attempt to justify the fact. If your attorney feels some answer of yours requires explanation, *he* may ask you when the other attorney has finished questioning you. (Most likely your attorney will ask no questions of you.)

4. *You should give only the information that you have readily at hand.* If you do not know certain information, do not attempt to give it. Do not turn to anyone in the room and ask for information. Do not promise more information and do not promise to look something up in the future unless your attorney advises it. If you have given the attorney representing you a document during the litigation, the contents of which document you cannot recall, and the opposing attorney asks for information contained on that document, simply state the truth—you do not know the answer. Make him ask you the proper question, which in this instance would be "Do you have any document or have you had any document that would contain the information in my previous question?" Do not voluntarily testify about the existence of documents possibly unknown to your opponent.

5. *Do not (a) reach into your pocket or briefcase for notes or memoranda, (b) produce any documents (particularly ones your attorney has not seen), (c) ask your attorney to produce documents you have furnished him, or (d) refer to any document in any way in answering any question unless the document is presented to you by the other side and approved by your attorney.* The purpose of a deposition is to find out the facts to which you can testify either of your own knowledge without documents or with the help of documents properly placed before you and identified on the record. Depositions are not a vehicle for the spontaneous production of documents.

6. *Do not answer a question that calls for a simple "yes" or "no" reply with a statement that invites further probing.* For example, you may be asked, "Did you speak with Mr. Smith last week?" If you answer, "No, I didn't speak with him *last* week," you are in actuality saying, "I spoke with him but not last week." Opposing counsel surely will continue questions concerning conversations with Mr. Smith. An answer of "No" is more likely to cut off further questioning, and it is all that you are required to give.

7. *Do not become angry or excited under any circumstances.* You may feel that the opposing attorney is rude, unnecessarily invading your privacy, or the like. If any behavior by the opposing attorney becomes unacceptable, your attorney shall handle the situation accordingly. Anger and excitement destroy the effect of testimony and you may inadvertently say things that, in either form or substance, may be used later to your disadvantage. Attorneys occasionally attempt to provoke anger in deponents in the hope that the deponent will make a testimonial error. Under no circumstances should you argue with the opposing attorney. Although sometimes difficult, attempt to give him the information in a courteous and unemotional tone of voice and manner. Displaying anger is precisely what he wants you to do.

8. *If your attorney begins to speak for any reason, stop your testimony immediately.* Do not speak again until he has completely finished speaking and listen carefully to what he says. If he objects to a question, do not answer the question unless instructed to do so. If instructed not to answer, do not answer under any circumstances;

do not argue with your attorney about whether you should answer, or suggest that you do not object to answering.

9. *Wait until the question is competed before beginning your answer.* (a) Although you may anticipate the balance of the question, you may guess incorrectly and, as a result, answer incorrectly. (b) Your attorney may want to object to the question. His objection is lost if you answer too quickly. (c) In order to accurately transcribe your testimony, the court reporter must hear the complete question and answer.

10. *You may take your time in answering a question.* Remember that the typed deposition will not show the length of time between the question and the answer. Of course, you should answer in a direct and straightforward manner, so as not to give the opposing counsel the impression that you are composing an answer. You should avoid being drawn into a serious of rapid-fire questions and answers by the other side. Remember that the opposing counsel is attempting to encourage you to "blurt out" answers before you have time to think or your attorney has time to object. Resist that temptation.

11. *If you are a client, in general, you are not required to give information that you learn in a conference with your attorney.* If you are asked a question that would require you to give such information, simply state that your answer would have to be based upon information learned from your attorneys and your attorney will make an appropriate objection, instructing you not to answer. (For expert witnesses, this is less so, but nevertheless follow your attorney's lead.)

12. *Freely admit discussing with counsel your probable deposition testimony.* Opposing counsel may ask "Have you previously discussed with anyone what your testimony will be?" and some witnesses, sensing an accusation of dishonesty, answer "No." Opposing counsel may then ask, "You mean you didn't even talk about the case with your attorney?" This puts you into the unsolvable bind of either (a) changing your sworn testimony, thereby creating the impression that you are unreliable, or (b) testifying falsely, which is wholly unacceptable. Tell the simple truth. Discussions between a client and an attorney concerning the facts in issue occur in every case; consultation is entirely proper and expected. Freely admit your preparation.

13. *Be very careful about estimating time or distance (or anything else).* Most people have difficulty with estimates. Examiners love for their witnesses to estimate. After an estimate you can expect a series of irritating questions attempting to require you to narrow down the range of your estimates. Obviously, that is a trap into which you should not fall.

14. *Under all circumstances, without exception, consistently and uniformly tell the truth.* The truth, whether in the deposition or on the witness stand at trial, will never really hurt a litigant. Your attorney can deal with the truth, but cannot deal with a fabrication. A lawyer may successfully "explain away" the truth, but no explanation exists for a witness' misstatement or concealment of the truth. In the eyes of the judge or jury, untruth devastates the credibility of a witness and hurts immeasurably. Also, as a practical mater, fabrications are difficult to recall with precision and can create opportunities for surprise. Tell the truth.

15. *Do not try to speculate before you answer whether a truthful answer will help or hinder your case.* Answer truthfully. Your answer should not vary because of the

effect you believe the answer will have on the case. The witness stand is a relatively poor place to make hurried judgments about the legal consequences of testimony. Avoid the temptation to adjust your answer in accordance with its possible consequences.

16. *Resist the temptation to give your side of the case in its entirety during deposition.* The time to present your case will come later—in a forum much more receptive than a deposition conducted by opposing counsel. After each question, consider the scope of the question and answer only the question you are asked. You need not elaborate, explain, or justify your answer.

17. *With respect to many issues in this litigation, your recollection has faded.* Generally, "not that I remember" or "not that I know of" is a better answer than "no" or the like. Avoid absolutes. Do not say "absolutely not," "never," "under no circumstances" or any other absolute, or irreversible response, unless it is truly called for.

18. *Watch out for leading questions such as "Isn't it true that" or "Isn't it more like."* The opposing lawyer is trying to either characterize your testimony by using terms favorable to his case or require a short answer to a question garnished with words that are argumentative or "loaded." If you have given a satisfactory answer, stick with it. You need not accept opposing counsel's characterizations or limit yourself to his vocabulary.

19. *Your knowledge necessarily includes hearsay.* The other lawyer may ask about hearsay during a deposition. In answering his questions, clearly distinguish between facts that are a matter of personal knowledge to you (things seen, heard, done, or spoken yourself) and matters that you know because you have been told them by someone else. This is an extremely important distinction.

20. *Never joke during a deposition.* The humor is not apparent on the typed transcript and you may look crude or cavalier about the truth. Avoid flippancy. Never use profanity—not even "hell" or "damn." Never use racist, sexist, ethnic, religious, or other slurs. (This is easy if you remember to conduct yourself during a deposition as if a jury were present.)

21. *Before, during, and after the deposition, do not chat with the opponents or the opposing attorney.* Remember, the other attorney and the opposing party are our legal "enemies." Do not let his friendly manner cause you to drop your guard.

22. *If you are handed a document by opposing counsel and asked a question about it, read the document before you answer the question.* Do not be concerned or nervous if the document is lengthy and requires 5, 10, or 15 minutes to read. Even if you think you recall the contents of the document, read it carefully in its entirety. If you insist (as you should) upon reading documents in their entirety, the opposing counsel will probably rephrase his question or otherwise avoid the necessity of your reading the document. However, if he does not rephrase or withdraw his question, read the document in its entirety.

23. *If you do not understand the question, insist that the opposing counsel rephrase the question.* If the question contains a premise with which you disagree or about which you have no knowledge, do not answer the question or answer it only after you have pointed out your disagreement or lack of knowledge.

Appendix 2

UNITED STATES
DEPARTMENT OF THE INTERIOR
BUREAU OF LAND MANAGEMENT
CALIFORNIA STATE OFFICE
2800 COTTAGE WAY
SACRAMENTO, CALIFORNIA 95825

January 30, 1980

TO WHOM IT MAY CONCERN:

I HEREBY CERTIFY That the attached reproductions are exact copies of documents on file in this office.

List of reproductions:

Field Notes of the Final Survey of the Rancho Las Pulgas, Traverse of the Salt Marsh from MA-8 through MA-13, surveyed by Thos. S. Stephens, Deputy Surveyor, under Instructions of October 25, 1856, and found in California Volume 479-32.

IN TESTIMONY WHEREOF I have hereunto subscribed my name and caused the seal of this office to be affixed on the above day and year.

Francis D. Eickbush
Chief, Branch of Cadastral Survey

Appendix 2

TABLE OF CASES

[References are to pages]

Dick v. New York Life Ins. Co., 359 U.S. 437,
79 S.Ct. 921 3 L.Ed.2d 935 (1959)

Dixon v. Pluns, 98 Cal. 384 (1893)

Dobson v. Whisenhant, 101 N.C. 645, 8 S.E.
126 (1888)

Doe v. Roe, 35 Del. 229, 162 A. 515 (1930)

Dolphin Lane Assoc. v. Town of Southampton,
372 N.Y.S.2d 52, 54 (1975)

Donnell v. Jones, 13 Ala. 490, 510 (1848)

Donnelly v. United States, 228 U.S. 708
(1913)

Donovan v. Fitzisimmons, 90 F.R.D. 583
(N.D. Ill. 1981)

Edmunds v. Atchinson etc. Ry. Co., 174 Cal.
246, 247, 162 P. 1038 (1917)

Elder v. Delcour, 269 S.W.2d 17, 23
(Mo. 1954)

Ellicott v. Pearl, 35 U.S. (10 Pet.) 412, 441
(1836)

Emmet v. Perry, 100 Me. 139, 141, 60 A. 872,
873 (1905)

Erie Railroad Co. v. Tompkins, 304 U.S. 64,
58 S.Ct. 817, 82 L.Ed. 1188 (1938)

Ernie v. Trinity Lutheran Church, 51 Cal.2d
702, 706 (1959)

Estate of Braue, 45 Cal.App.2d 502, 505,
114 P.2d 386 (1941)

Emerson, 170 Cal. 81, 82, 148 P. 523 (1915)
Gains, 15 Cal.2d 255 (1940)
Galvin, 114 Cal.App.2d 354, 360, 250 P.2d
333 (1952)
LeSure, 21 Cal.App.2d 73, 83, 68 P.2d 313
(1937)
McConnell, 6 Cal.2d 493, 499, 58 P.2d
639 (1936)
Pieper, 224 Cal.App.2d 670, 691 (1964)
Pitcairn, 6 Cal.2d 730, 733, 59 P.2d 90
(1936)

*Ethyl Corp. v. Environmental Protection
Agency,* 478 F.2d 47 (4th Cir. 1973)

Evans v. Merthyr Tydfil, 1 Ch. 241 (1899)

Faekler v. Wright, 86 Cal. 217, 24 P. 996
(1890)

FCC v. Schreiber, 381 U.S. 279 (1965)

Ferris v. Emmons, 214 Cal. 501 (1931)

Finberg v. Gilbert, 104 Tex. 539, 141 S.W.
82 (1911)

Foley v. Foley, 120 Cal. 33 (1898)

Foss v. Johnstone, 158 Cal. 119, 110 P. 294
(1910)

Frye v. United States, 54 App.D.C. 46, 47,
293 F. 1013, 1014 (D.C. Cir. 1923).

Gagnon Co. v. Nevada Desert Inn, 45 Cal.2d
448, 289 P.2d 466 (1955)

Galloway v. United States, 319 U.S. 372
(1943)

Garner v. Wofinbarger, 430 F.2d 1093
(5th Cir. 1970)

Georgia v. South Carolina, 257 U.S. 516
(1922)

Gibbons v. Ogden, 22 U.S. (9 Wheat.) 1
(1824)

Gibson v. Poor, 21 N.H. 440, 53 Am. Dec.
216 (1850)

Gion v. City of Santa Cruz, 2 Cal.3d 29, 39,
44 (1970)

Glantz v. Freedman, 100 Cal.App. 611
(1929)

Goyette v. Keenan, 196 Mass. 416, 82 N.E.
427 (1907)

Gray v. Reclamation Dist. No. 1500, 174 Cal.
622 (1917)

Gray v. Robinson, 33 Cal.App.2d 177, 91 P.2d
194 (1939)

Guardianship of Levy, 137 Cal.App.2d 237
(1955)

Guyer v. Snyder, 133 Md. 19, 104 A. 116
(1918)

*Hammond Lumber Co. v. County of Los
Angeles,* 104 Cal.App. 235 (1930)

*Hancock v. Supreme Council Catholic
Benevolent Legion,* 69 N.J.L. 308, 55 A.
246 (1903)

Hans v. Louisiana, 134 U.S. 1 (1890)

Hansen v. G & G Trucking Co., 236 Cal.
App.2d 481, 46 C.R. 186 (1965)

Hardin v. Jordan, 140 U.S. 371, 380–81
(1891)

Harmon Enterprises, Inc. v. Vroman,
167 Cal.App.2d 517 (1959)

Harriman v. Brown, 35 Va. (8 Leigh) 697,
707 (1837)

Harrington v. Goldsmith, 136 Cal. 168
(1902)

Harris v. United States, 413 F.2d 316
(9th Cir. 1969)

A

About the Author

Mr. Briscoe has been a member of the San Francisco-based law firm Washburn, Briscoe & McCarthy since 1980. From 1972 through 1980 he had been a deputy attorney general of the State of California. His legal practice has concentrated on natural resources, environmental,

(*Continued*)

and real property law. He received his law degree from the University of San Francisco in 1972, where he graduated with honors and was a member of the Law Review.

He has tried civil cases in the state and federal courts, and has argued before the California Supreme Court and the United States Supreme Court. Since 1982, Mr. Briscoe has served as special counsel to the State of Alaska in Supreme Court litigation against the United States concerning the limit of American territorial waters off the north coast of Alaska, between Icy Cape and the United States-Canadian boundary. Mr. Briscoe has also served as special counsel to the State of Georgia in its boundary dispute with the State of South Carolina, and as special assistant attorney general for the State of Hawaii for the purpose of assessing that State's claims to the resources of the submerged lands of the Hawaiian archipelago. He is as well special counsel to the Territory of Guam in various land matters.

Mr. Briscoe serves as special adviser to the United Nations Compensation Commission in Geneva. The Commission is hearing the war-reparations claims arising from the Iraqi invasion and occupation of Kuwait. Mr. Briscoe is advising the Commission regarding the environmental and natural resource damages that occurred as a result of the oil spills and fires in Kuwait in 1991.

Since 1990 Mr. Briscoe has been a Visiting Scholar at the Boalt Hall School of Law at the University of California, Berkeley.

Postscript

It was suggested in the Introduction that this book might serve to infect the reader with the conviction, which Melville pretended to share, that there is reason, even in the law. Having read some or all of this volume, the reader might find himself, however, more inclined toward the sentiments of Mr. Justice Jackson of the United States Supreme Court, who in 1947 dissented from the Court's decision in *Securities Exchange Commission v. Chenery Corp.*, 332 U.S. 194, 213–214, and wrote:

> The Court's reasoning adds up to this: the Commission must be sustained because of its accumulated experience in solving a problem with which it has never been confronted! . . . I give up. Now I realize what Mark Twain meant when he said, 'The more you explain it, the more I don't understand it.'

Postscript

Printed and bound by CPI Group (UK) Ltd, Croydon, CR0 4YY

27/10/2024

14580314-0003